Advanced Fossil Power Systems Comparison Study

Final Report

Prepared for:
National Energy Technology Laboratory
P.O. Box 880, 3610 Collins Ferry Road
Morgantown, WV 26507-0880
and
P.O. Box 10940, 626 Cochrans Mill Road
Pittsburgh, PA 15236-0940

Prepared by:
Edward L. Parsons
National Energy Technology Laboratory
P. O. Box 880, 3610 Collins Ferry Road
Morgantown, WV 26507-0880

Walter W. Shelton
Jennifer L. Lyons
EG&G Technical Services, Inc.
3604 Collins Ferry Road Suite 200
Morgantown, West Virginia 26505

December, 2002

DISCLAIMER

This report was prepared as an account of work sponsored by an agency of the United States Government. Neither the United States Government nor any agency thereof, nor any of their employees, makes any warranty, express or implied, or assumes any legal liability or responsibility for the accuracy, completeness, or usefulness of any information, apparatus, product, or process disclosed, or represents that its use would not infringe privately owned rights. Reference herein to any specific commercial product, process, or service by trade name, trademark, manufacturer, or otherwise does not necessarily constitute or imply its endorsement, recommendation, or favoring by the United States Government or any agency thereof. The views and opinions of authors expressed herein do not necessarily state or reflect those of the United States Government or any agency thereof.

TABLE OF CONTENTS

TABLE OF CONTENTS (continued)

TABLE OF CONTENTS (continued)

Tables

Figures

Figures (continued)

Advanced Fossil Power Systems
Comparison Study

SUMMARY

Aspen Plus® (version 10.2) simulation models and the Cost of Electricity (COE) have been developed for advanced fossil power generation systems both with and without carbon dioxide (CO_2) capture. The intent was to compare the cycles based on using common assumptions and analytic standards with respect to realizable performance, cost, emissions and footprint. Additionally, commercially available (or near term) reference plants were included for comparison.

The advanced fossil power systems considered were: (both natural gas and coal fueled)
- Hydraulic Air Compression Cycle (HAC)
- Rocket Engine Gas Generator Cycle
- Hydrogen Turbine (air) Cycle
- Hybrid Cycle (Turbine / Solid-Oxide Fuel Cell)
- Humid Air Turbine Cycle (HAT) [(CO_2) capture – not considered]

Reference Plants developed based on previous NETL/EG&G studies included:
- Pulverized Coal (PC) Boiler
- Natural Gas Combined Cycle (NGCC)
- Integrated Gasification Combined Cycle (IGCC)

Capital cost estimates were developed for the above cases using data from the EG&G Cost Estimating Notebook (version 1.11) and several contractor reports. The format follows the guidelines set by EPRI TAG methods. Individual equipment sections were based on capacity factored techniques. The costs are reported in first quarter 2002 dollars. The total capital requirement includes equipment, labor, engineering fees, contingencies, interest during construction, startup costs, working capital and land. Other assumptions are provided in summary tables in Appendix B which contains the COE spreadsheets developed for all cases.

Results are compared in Table 1 (Natural Gas Cycles) and in Table 2 (Coal Cycles). These results demonstrate the following key observations:

- For all systems, (CO_2) capture entails major cost & efficiency penalties.
- Only Hybrids perform at or near the Vision 21 efficiency goals summarized in Appendix D.
- Rocket Engine cycles have lower efficiency and higher cost than other options requiring far less development.
- HAC cycles based on a closed-loop water system are unattractive. An open-loop water system (dam site) may be attractive as a niche market.
- Hydrogen Turbine (air) and HAT cycles are also unattractive.

TABLE 1 - Natural Gas Cycles

POWER SYSTEM / Power Generation Cycle	NATURAL GAS COMBINED CYCLE (NGCC)		HYDRAULIC AIR COMPRESSION (HAC)		ROCKET ENGINE (CES)	HYDROGEN TURBINE (HT)	Hybrid Cycle	HUMID AIR TURBINE (HAT)
	NGCC "G" Gas Turbine	NGCC "G" Gas Turbine (CO2 Capture)	HAC NATURAL GAS	HAC NATUTAL GAS (CO2 CAPTURE)	CES (gas generator) (CO2 CAPTURE)	HT (H2 FROM SMR) (CO2 CAPTURE)	Hybrid Turbine (Siemens/West.) -SOFC / Turbine	HAT (PW GT) Natural Gas
Net Power MWe	379.1	326.9	323.5	300.2	398.4	413.1	19	318.7
Net Plant Efficiency %LHV	57.9	49.9	53.2	43.8	48.3	64.4 (H_2) / 42.9 (NG)	67.3	57.6
Total Capital Requirement $ / KW	515	911	681	1140	975	1323	1476	873
Cost of Electricity $ / MW-hr	34.7	48.3	44.2	61.0	49.2	63.5	53.4	47
NOx emissions lb/MW-hr	0.176	0.204	0.194	0.210	NEG	0.161	0.0132	0.074
Sox emissions lb/MW-hr	—	—	—	—	—	—	—	—
CO2 Production lb/MW-hr								
a) Emitted to atmosphere	757	88	824	100		*	661	758
b) Sequesterable		790		899	901	719		
Footprint (battery limits) sq ft/MW	282	362	179	230	825	472	1120	175

2

Table 2 - Coal Cycles

POWER SYSTEM	PULVERIZED COAL (PC)			INTEGRATED GASIFICATION COMBINED CYCLE (IGCC)				
Generation Cycle	PC Steam Cycle (no CO2 Capture)	PC Steam Cycle (amine CO2 Capture)	PC Steam Cycle (O2 Boiler/ CO2 CAPTURE)	IGCC Destec (E-Gas) CGCU "G" Gas Turbine	IGCC Destec (E-Gas) HGCU "G" Gas Turbine	IGCC Destec (E-Gas) CGCU "G" Gas Turbine (CO2 Capture)	IGCC SHELL CGCU "G" Gas Turbine	IGCC SHELL CGCU Gas Turb (ANL) (CO2 Capture)
Net Power MWe	396.8	283	298.4	400.6	400.4	358.6	412.8	351.1
Net Plant Efficiency % LHV	38.9	27.7	30.5	46.7	49.4	40.1	47.4	40.1
Total Capital Requirement $ / KW	1268	2373	2259	1374	1354	1897	1370	2270
Cost of Electricity $ / MW-hr	42.3	76.6	68.8	40.9	39.1	54.4	40.6	62.9
NOx emissions lb/MW-hr	4.09	5.74	0.27	0.165	0.165	0.185	0.160	0.182
Sox emissions lb/MW-hr	3.12	4.38	3.97	0.342	0.04	0.113	0.276	0.112
CO2 Production lb/MW-hr								
a) Emitted to atmosphere	1837	129	*	1517	1431	231	1496	190
b) Sequesterable		2448	2332			1536		1569
Footprint (battery limits) sq ft/MW	636	1009	1591	1092	1057	1198	1065	1168**

Table 2 - Coal Cycles (continued)

POWER SYSTEM	HYDRAULIC AIR COMPRESSION (HAC)		ROCKET ENGINE (CES)	HYDROGEN TURBINE (HT)	HYBRID CYCLE (HYB)			HUMID AIR TURBINE (HAT)
Generation Cycle	HAC Destec (E-Gas) CGCU	HAC Destec HP (E-Gas) HGCU (CO2 CAPTURE)	CES (gas generator) Destec HP (E-Gas) HGCU (CO2 CAPTURE)	HT Destec HP (E-Gas) HGCU (CO2 CAPTURE)	HYB Destec (E-Gas) HGCU "G" GT / SOFC (NO CO2 CAPTURE)	HYB Destec HP (E-Gas) HGCU/HSD "G" GT / SOFC (CO2 CAPTURE)	HYB Destec (E-Gas) OTM / CGCU "G" GT / SOFC (NO CO2 CAPTURE)	HAT (PW GT) Destec (E-Gas) CGCU
Net Power MWe	325.9	312.4	406.2	375.3	643.6	754.6	675.2	407.4
Net Plant Efficiency % LHV	43.8	35.2	41.4	38	56.4	49.7	57	44.9
Total Capital Requirement $ / KW	1436	2189	1768	1909	1508	1822	1340	1411
Cost of Electricity $ / MW-hr	47.0	65.5	49.3	53.6	41.1	48.8	38	42.1
NOx emissions lb/MW-hr	0.193	0.204	NEG	0.177	0.107	0.093	0.101	0.071
Sox emissions lb/MW-hr	0.337	0.048	0.044	0.046	0.005	0.004	0.014	0.353
CO2 Production lb/MW-hr								
a) Emitted to atmosphere	1561	142		131	1254	101	1237	1576
b) Sequesterable		1870	1702	1731		1323		
Footprint (battery limits) sq ft/MW	1293	1583	1458	1445	1310	1408	1388	811

I. REFERENCE PLANTS

I-1 PULVERIZED COAL (PC) BOILER

PC Boiler power plants without CO_2 capture represent a large number of the existing coal-fired power plants used for generating electrical power in the United States and North America. Three cases were developed based on previous Aspen Plus® simulations [1] for use as reference plants to contrast performance and cost with proposed advanced fossil power systems. The first case (Base Case) represents a modern power plant that employs both particulate and sulfur recovery. The remaining two cases are variations that add the possibility of CO_2 capture. The Base Case is an air-blown 400 MWe power plant without CO_2 capture that is used to establish baseline power plant performance and to assess the cost of electricity (COE). In the second case, an amine absorption process is added to capture CO_2 from the flue gas. The third case replaces the air used in the PC base case with a mixture of oxygen and recycled flue gas as the oxidant stream sent to the PC Boiler. This results in a flue gas stream containing primarily CO_2 and water vapor. Water is separated by condensation from the flue gas portion that is not recycled to obtain a concentrated CO_2 stream for sequestration. In both cases that capture CO_2, the CO_2 - rich stream was compressed to 1500 psia and leaves as a high pressure gas stream. (Further compression to approximately 2100 psia would be required to obtain a liquid stream. This would lower the process efficiency and raise the COE somewhat compared to the values listed in this report).

For the two cases with CO_2 capture, the boiler capacity was chosen the same as the base case to maintain the steam generation at the same amount. Any power or steam required for the CO_2 capture or the cryogenic oxygen plant was imported internally from the power plant. As a result, the net power production was reduced. It should be stressed that PC Boiler plants with CO_2 capture as described in these two cases are technically possible but are not currently existing commercial units due to both efficiency and cost penalties.

I-1.1 PC Power Plant - Base Case – Description

The Base Case consists of a power plant based on a pulverized coal (PC) boiler and steam turbine. The system described in a report by Buchanan et al. [2] was used as a design basis. This case was evaluated for benchmarking the performance of the other cases. A single reheat steam power cycle (2400psig/1000 °F /1000 °F) was used to generate 400 MWe of power. The steam generator was a natural circulation, wall-fired, subcritical unit arranged with a water-cooled dry-bottom furnace, superheater, reheater, economizer and air heater. The burners were low-NOx type. The flue gas was desulfurized by scrubbing with lime slurry. A simplified flow diagram is shown in Figure 1.

In this process, air is preheated in an air heater by exchanging heat with the flue gas. Coal and hot air are fed to the boiler from the bottom. High pressure steam is generated in the radiant section. Flue gas from the radiant section enters the convective section at 2200 °F. In the convective section, thermal energy from the flue gas is transferred to high-pressure steam, intermediate pressure steam and feed water. Flue gas leaves the convective section at 600 °F and passes through the air heater to preheat air. A precipitator is used to remove particulates and the flue gas is then sent to a SO₂ scrubber with the aid of an induced draft fan. Lime slurry is employed to scrub SO₂ from the flue gas. The cleaned flue gas leaves through the stacks. The high-pressure steam is superheated in the convective section. Superheated steam at 2415 psia and 1000 °F is expanded in the high-pressure turbine to an intermediate pressure of 604 psia. This IP steam is reheated in the convective section to 1000 °F and is then expanded in the IP steam turbine. Finally, the exhaust from the IP steam turbine is expanded in the LP (low pressure) turbine to 1 psia and enters the condenser. The condensate water is sent to a series of low-pressure feed heaters. The heated water is sent to the deaerator to remove dissolved gases. Deaerated water is passed through the high-pressure water heaters and is then fed to the economizer portion of the boiler's convective section. Water is further heated to close to its saturation temperature in the economizer and then sent to radiant section for boiling.

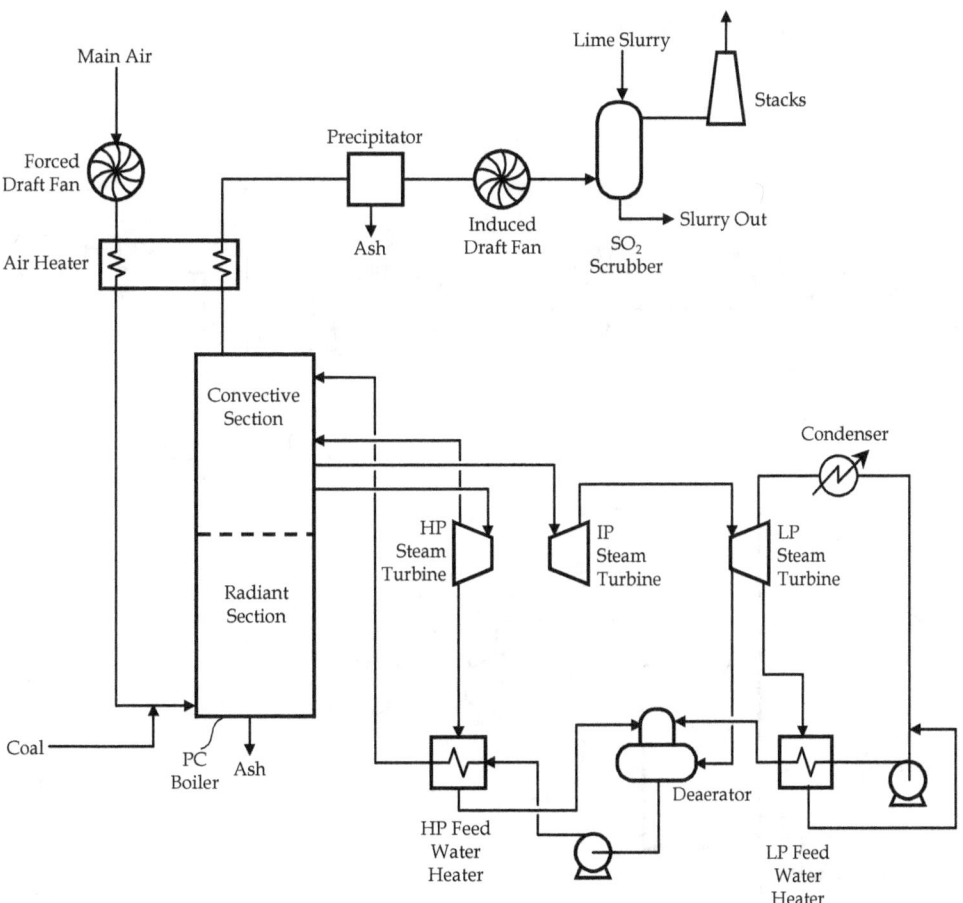

Figure 1. Pulverized Coal Boiler Power Plant

I-1.2   PC Power Plant - Amine CO_2 Capture – Description

In this case, the boiler operation is identical to the base case; i.e. air is used as the oxidant. The flue gas after sulfur removal is sent to an amine plant for CO_2 separation. In the amine plant, a MEA based solution is used to absorb CO_2 from the flue gas. The CO_2-depleted gas from the absorber is vented to the atmosphere. The CO_2-rich solvent is heated by lean solvent and then sent to a stripper for regeneration. Low-pressure steam (35 psia) is extracted from the LP turbine section and sent to the stripper reboiler of the amine plant. A concentrated CO_2 stream is recovered from the stripper and the lean solvent is recycled to the absorber. The CO_2 stream is compressed to1500 psia in a multistage intercooled compression section and leaves as a high pressure gas. The condensed water from the stripper reboiler is sent back to the steam cycle. Extraction of steam reduces significantly the gross power output from the steam turbines. Additionally, the amine plant consumes power for the flue gas blower and for the amine solvent recirculation pumps and a large power consumption is due to the required CO_2 compressor.

A simplified flow diagram is shown in Figure 2.

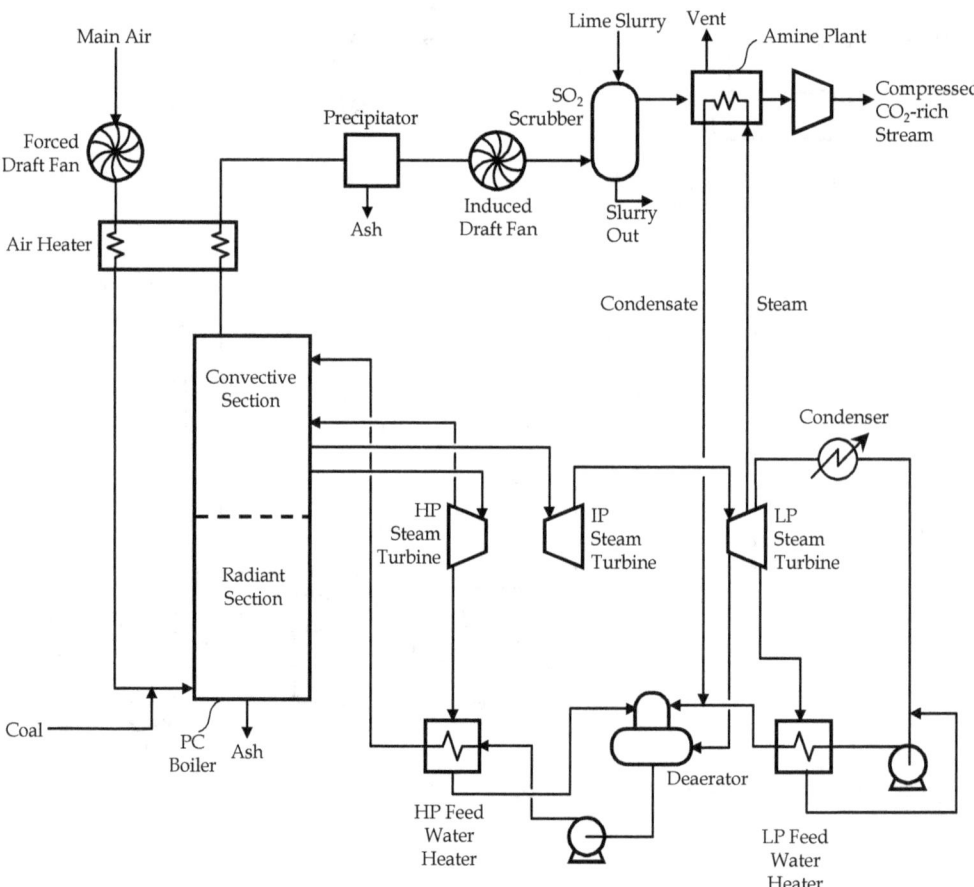

Figure 2. PC-Fired Boiler with Amine Scrubbing for CO_2 Sequestration

I-1.3 PC Power Plant – Cryogenic ASU – Description

A cryogenic ASU supplies oxygen to the PC fired boiler. Oxygen with 95% purity was selected, because the cost of oxygen is significantly lower than that for high-purity oxygen (>99.5% purity). A portion of the flue gas is recycled and mixed with oxygen from the cryogenic ASU. The resulting oxidant stream (mixture of O_2, CO_2 and H_2O and small amounts of Ar and N_2) is preheated in the inlet heater and fed to the boiler along with pulverized coal. Since most of the nitrogen from air is eliminated in the ASU, the flue gas leaving the boiler essentially contains CO_2 and water vapor. After the flue gas preheats the oxidant stream, it passes through a precipitator and the portion that is not recycled enters the SO_2 scrubber. Water is condensed out of the flue gas stream exiting the scrubber and a concentrated CO_2 stream is obtained. The CO_2-rich stream is compressed to 1500 psia for sequestration.

This case was iterated by adjusting flue gas recycle flow, oxygen flow and coal flow. The goal was to achieve the same temperatures for flue gas leaving the radiant and convective sections as those in the base case and to generate the same amount of steam from the boiler as the base case. Overall, the power generated from steam turbines was roughly the same as in the base case. However, a significant portion of the power is supplied to the ASU and the CO_2 compressor. A simplified flow diagram is shown in Figure 3.

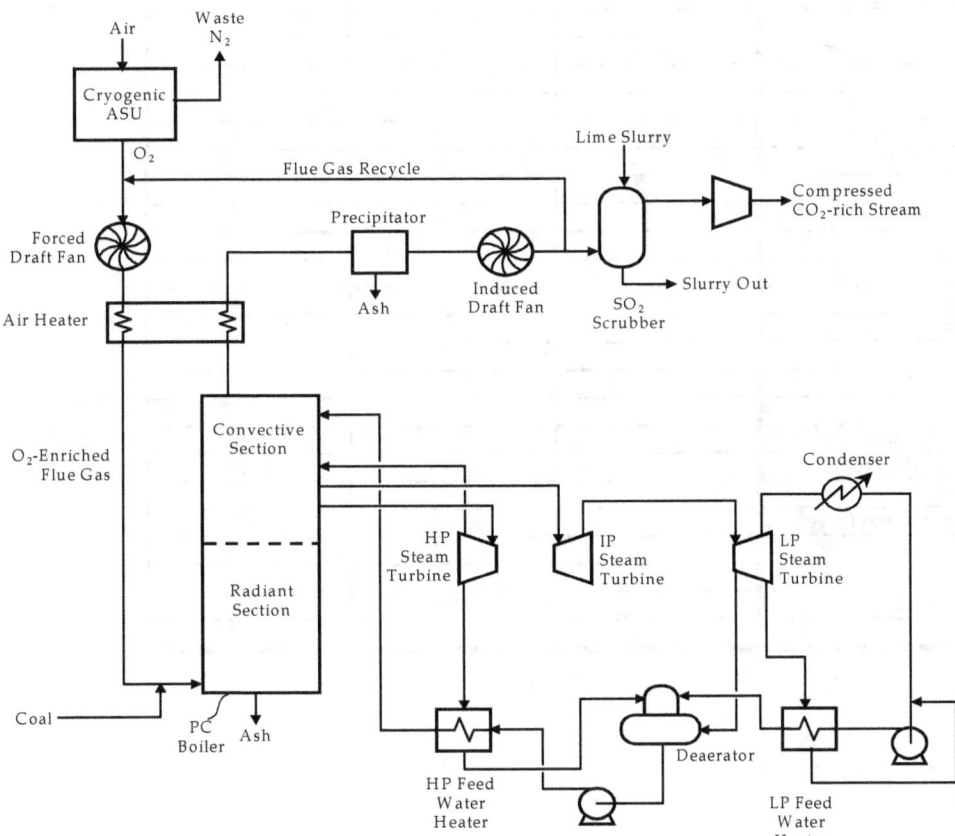

Figure 3. PC Fired Boiler with Flue Gas Recycle for CO_2 Sequestration; O_2 from Cryogenic ASU

I-1.4 PC Power Plant Results

Detailed flow diagrams with stream summaries are provided in Appendix A based on the Aspen Plus® simulation results. Emissions for NOx and SOx were based on the BACT (best available control technology) and CO_2 was based on simulation results. Capital cost estimates were developed based on Buchanan et al. [2] and vendor estimates for the amine plant and the oxygen plant [3]. Spreadsheets showing capital costs and the COE analysis are provided in Appendix B. The results shown below for these cases illustrate significant cost and efficiency penalties for CO_2 capture.

Table 3. Pulverized Coal (PC)

POWER SYSTEM	PULVERIZED COAL (PC)		
Generation Cycle	Coal PC Steam Cycle (no CO2 Capture)	Coal PC Steam Cycle (amine CO2 Capture	Coal PC Steam Cycle (O2 Boiler/ CO2 CAPTURE)
Net Power MWe	396.8	283	298.4
Net Plant Efficiency % LHV	38.86	27.72	30.5
Total Capital Requirement $ / KW	1268	2373	2259
Cost of Electricity Constant $ / MW-hr	42.3	76.6	68.8
NOx emissions lb/MW-hr	4.09	5.74	0.205
Sox emissions lb/MW-hr	3.12	4.16	2.98
CO2 Production lb/MW-hr			
a) Emitted to atmosphere	1837	129	*
b) Sequesterable		2448	2332
CO2 concentration (mole%) (in sequestered gas)		99.70%	86.60%
Footprint (battery limits) sq ft/MW	636	1009	1591

The Base Case power plant generates 396.8 MW and its efficiency is 38.9% (LHV) or 37.5% (HHV). The CO_2 capture decreases the efficiency by a dramatic 8 – 11 percentage points and and nearly doubles the base case's total capital requirement of $1268/KW.

The cost and performance of the amine plant are based on commercially available oxygen-tolerant amine technology designed to capture 95% of the CO_2. The energy consumption for the amine case was assumed to be 3.7 MMBtu / ton CO_2 recovered. (NETL is currently funding research aimed at reducing this by up to 50% , [4]). Steam consumption for regenerating the amine solution resulted in a significant penalty on power production. The power output from the steam turbine decreased to 325 MW. The consumption of power by the amine plant and the CO_2 compressor reduced the net power output from the power plant to 283 MW. Thus, 114 MW power was consumed for the CO_2 capture system. Overall efficiency of the system was 27.7% (LHV). Based on vendor information, the amine plant and CO_2 compression added $122 MM in capital cost to the base case. This increased the COE from 42.3 to 76.6 ($/MW-hr, Constant $ basis).

In the last case, PC oxygen/recycle flue gas boiler, it was assumed that the concentrated CO_2 stream can be sequestered without further processing. Thus, the entire CO_2-rich flue gas stream (not recycled) was compressed to 1500 psia for sequestration and there were no CO_2 emissions in this case. The cryogenic ASU produced 7570 tpd oxygen (on pure basis) of 95% purity (by vol.) and consumed 64 MW power. The compression of the CO_2-rich stream consumed another 34 MW. Use of oxygen increased the boiler efficiency as evidenced by reduced coal consumption. However, the net power output for the cryogenic case decreased to 298 MW and the efficiency decreased to 29.5%. Additional capital cost of $145 MM included the cost of the cryogenic ASU, the cost of redesigning the normal PC boiler for oxygen firing and the capital cost of the CO_2 compressor. The COE with CO_2 capture was $68.8/ MW-hr.

I-2 NATURAL GAS COMBINED CYCLE (NGCC)

Aspen Plus® simulations were developed for two natural gas power combined cycle power plants using a gas turbine model that is based on the Siemens-Westinghouse W501G gas turbine and a three pressure level steam cycle. The two cases differ depending on whether CO_2 capture is included. The first case (no CO_2 capture) produces 379.1 MWe at a process efficiency of 57.9% (LHV) and is considered as a commercially available plant. The second case includes CO_2 capture based on recovering CO_2 from the flue gas stream that exits the heat recovery steam generator (HRSG). The CO_2 capture envisioned is based on a commercial amine process (Dow Chemical) [5] operating at a design of 90% CO_2 capture coupled with compression to sequester the CO_2 as a high pressure liquid. The power is reduced both due to compression and the steam required for regenerating the amine solvent. Dow Chemical has advised us that the system is both more difficult when compared with recovery from a PC power plant and more expensive due to the higher oxygen content in the exhaust. At the present time, they were unaware of any existing plant using this approach due to the high efficiency penalty expected.
The Aspen Plus® results indicated a reduction in power to 326.9 MWe and a reduction in efficiency to 49.9% (LHV). Results are summarized in the following table.

Table 4. Natural Gas Combined Results

POWER SYSTEM	NATURAL GAS COMBINED CYCLE (NGCC)	
Power Generation Cycle	NGCC "G" Gas Turbine	NGCC "G" Gas Turbine (CO2 Capture)
Net Power MWe	379.1	326.9
Net Plant Efficiency % LHV	57.9	49.9
Total Capital Requirement $ / KW	515	911
Cost of Electricity $ / MW-hr	34.7	48.3
NOx emissions lb/MW-hr	0.176	0.204
Sox emissions lb/MW-hr	---	---
CO2 Production lb/MW-hr		
a) Emitted to atmosphere	757	88
b) Sequesterable		790
Footprint (battery limits) sq ft/MW	282	362

I-2.1 NGCC – No CO_2 Capture

This power cycle is considered to be commercially available. The gas turbine conditions [6] (see "Gas Turbine World" - Siemens-Westinghouse W501G) used were:

Pressure Ratio:	19.2 : 1
Inlet Air Flowrate :	1241 lbs/sec
Exhaust Temperature:	1101 $^{\circ}$F
Turbine Inlet Temperature:	2583 $^{\circ}$F

The Steam Cycle was based on a heat recovery steam generation (HRSG) section that generates steam at three pressure levels with power recovered in a steam turbine system using a single reheat and at conditions: 1800 psia / 1000 $^{\circ}$F / 492 psia / 1000 $^{\circ}$F.

Emissions were based on simulation results for CO_2 and an assumed NOx level of 9 ppmv. (the table results would be slightly higher if adjusted for 15% oxygen level in the exhaust – which is often given in reports).

The capital cost estimate was based on information published in NETL reports , DOE/HQ contractor studies and from the Gas Turbine World (2001) annual summary [6]. The cost of electricity analysis was based on the EPRI Tag method.

The Footprint (battery limits) was a crude estimate based on available information in published studies (such as the footprint of the W501G gas turbine). The actual plant site would be approximately 100 acres.

In Figure 4, the process is shown with key process streams to illustrate this power plant cycle. Appendix A contains detailed information for the process streams shown.

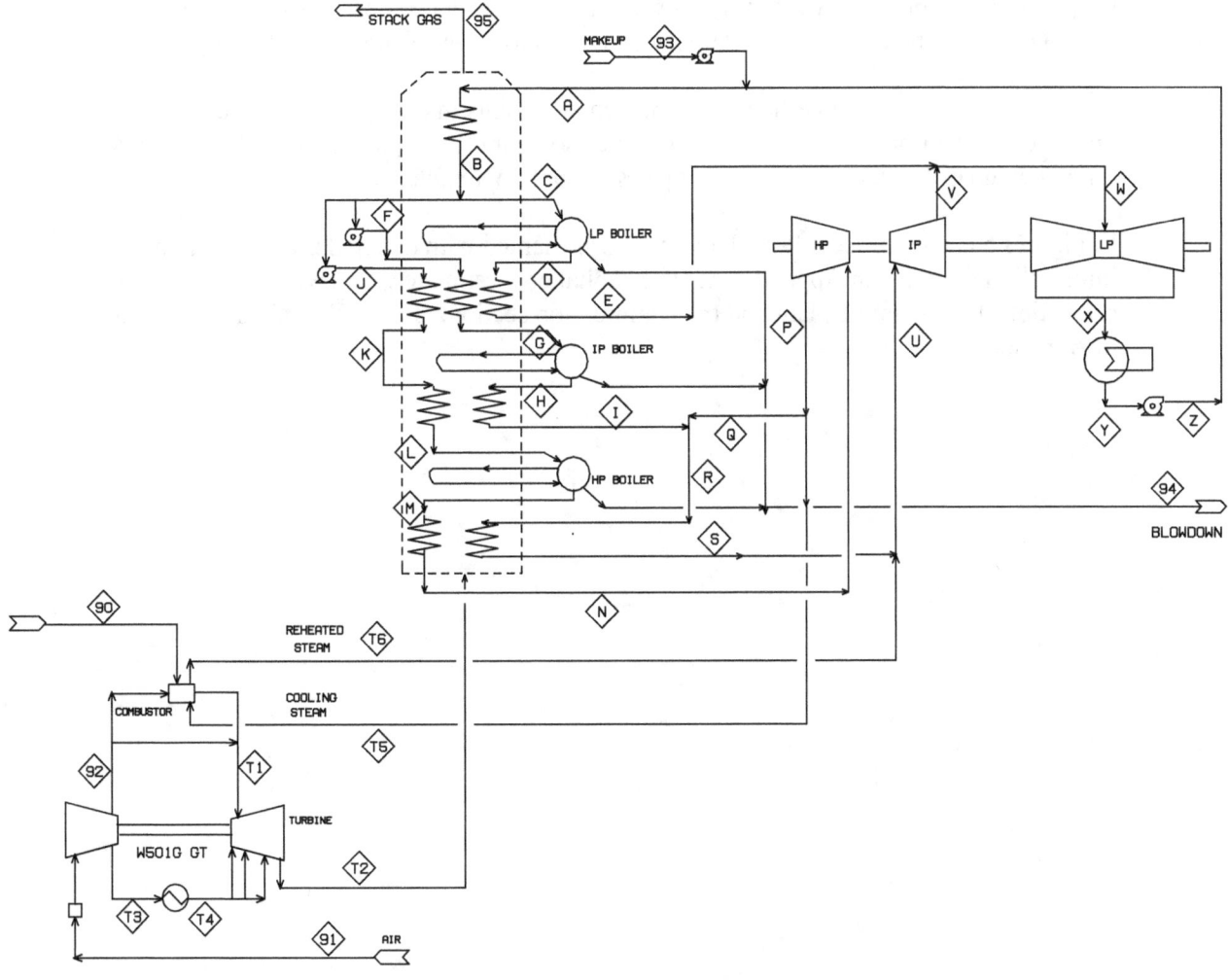

Figure 4. NGCC Power Plant

I-2.2   NGCC – CO_2 Capture

An Aspen Plus$^®$ simulation was developed based on adding a CO_2 capture process. This was accomplished by adding an amine plant followed by a compression section to the previous case. Figure 5 and Figure 6 show the modifications.

The flue gas exiting the HRSG enters an amine plant shown in Figure 6 to produce a CO_2 rich-stream. This stream is compressed in an inter-cooled five stage compressor to a pressure of 2160 psia. The high pressure CO_2 gas stream is cooled to approximately 100 ^{o}F to produce a liquid stream which is pumped to 3000 psia to complete the CO_2 capture. The system simulated used a design basis of 90% CO_2 capture and an energy input for the reboiler in the amine plant of 3.7 MMBtu / ton CO_2 recovered. This energy requirement is met by low pressure steam (35 psia) which is withdrawn from the steam cycle prior to the low pressure steam turbine. (see Figure 5). This results in a loss of power in the steam cycle and when combined with the compression power requirement results in a significant power penalty for CO_2 capture. Table 4 above shows that the net power produced decreases to 326.9 MWe from 379 MWe and the overall efficiency decreases to 49.9% from 57.9% (LHV).

Even when an increase of perhaps 4 – 6 percentage points in efficiency is added for an improved ATS turbine system and an improved solvent process, the Vision 21 program's efficiency goals for natural gas power cycles are not obtainable.

In Figures 5 and 6, process flow diagrams are presented with detailed process stream information provided in Appendix A. The capital cost estimate was developed by adding projections for the amine plant and the compression section. The COE results are provided in Appendix B.

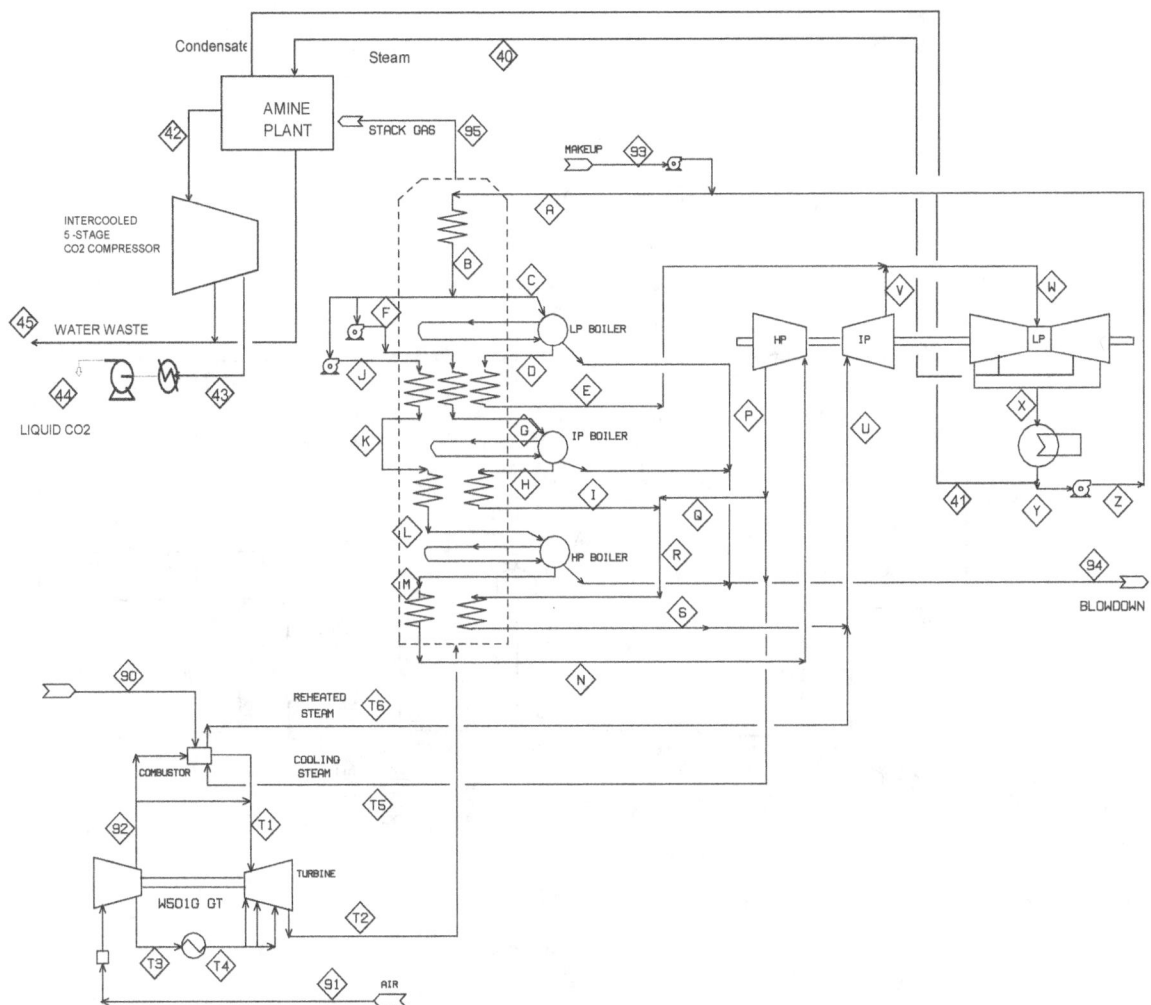

NGCC (WITH CO2 CAPTURE) - W501G GAS TURBINE - 3 PRESSURE LEVEL STEAM CYCLE

Figure 5. NGCC – with CO$_2$ Capture

AMINE PLANT

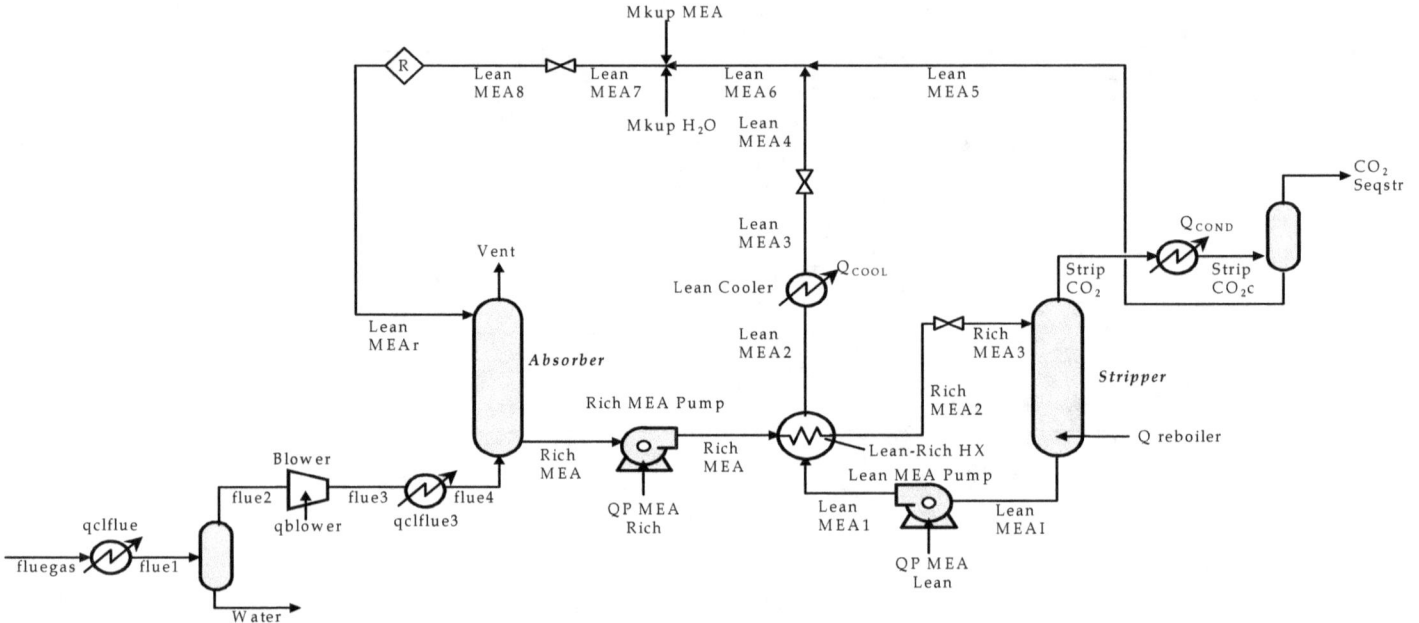

Figure 6. Amine Plant

I-3. INTEGRATED GASIFCATION COMBINED CYCLE (IGCC)

NETL/DOE has been sponsoring the research and development of IGCC as the cleanest coal-based power system available today for several decades and in a recent report (July 2002) [7] a snapshot is provided from industry's viewpoint on the outlook and needs for future research and development of both IGCC and Gasification Technologies. As part of providing a comparison with the proposed advanced coal power systems presented later in this report, a group of IGCC systems studies has been assembled based on previous NETL studies completed in FY2000. In Table 5, results are summarized for several reference IGCC cases that are viewed as near-term commercially available and for a case proposed on the inclusion of a hydrogen powered fuel cell. (These systems studies are available with additional systems based on different gasifiers on the NETL website [8].) Key assumptions include:

- Oxygen- blown Gasification (Destec [E-Gas™] or Shell) using Illinois No. 6 bituminous coal.
- Gas Cleanup for particulate matter, chloride and sulfur based on either Cold Gas Cleanup or Hot Gas Cleanup.
- Gas Turbine based on Siemens Westinghouse W501G heavy duty gas turbine with dry low-NOx combustor. (9 ppmv NOx, nominal 272 MWe – modified for syngas).
- Steam Cycle is a three pressure level process.
- Air Separation based on cryogenic process integrated with the gas turbine.
- Single-Train IGCC Power Plants.
- For the two cases that include CO_2 sequestration, the CO_2 is captured and compressed to provide a liquid product stream.
- For the case that produces high purity hydrogen, conversion to power via a fuel cell occurs at 65% of the heating value of the hydrogen produced.
- Cost of Electricity (COE) based on estimates updated to First Quarter 2002 ,

These cases demonstrate overall efficiencies (LHV basis) ranging from 40- 49%. The lower efficiencies cases include a CO_2 Sequestration penalty of 6 – 7 percentage points.

Table 5. Reference IGCC Case Results

POWER SYSTEM	INTEGRATED GASIFICATION COMBINED CYCLE (IGCC)				
	Case 1	Case 2	Case 3	Case 4	Case 5
Generation Cycle	IGCC Destec (E-Gas) CGCU "G" Gas Turbine	IGCC Destec (E-Gas) HGCU "G" Gas Turbine	IGCC Destec (E-Gas) CGCU "G" Gas Turbine (CO2 Capture)	IGCC SHELL CGCU "G" Gas Turbine	IGCC SHELL CGCU Gas Turb (ANL) (CO2 Capture)
Net Power MWe	400.6	400.4	358.6	412.8	351.1
Net Plant Efficiency % LHV	46.7	49.4	40.1	47.4	40.1
Total Capital Requirement $ / KW	1374	1354	1897	1370	2270
Cost of Electricity Constant $ / MW-hr	40.9	39.1	54.4	40.6	62.9
NOx emissions lb/MW-hr	0.165	0.165	0.185	0.160	0.182
Sox emissions lb/MW-hr	0.342	0.04	0.113	0.276	0.112
CO2 Production lb/MW-hr					
a) Emitted to atmosphere	1517	1431	231	1496	190
b) Sequesterable			1536		1569
Footprint (battery limits)	1092	1057	1198	1065	1168**

(** Footprint does not include fuel cell)

I-3.1 IGCC Destec (E-Gas™) Cases – No CO$_2$ Capture

Two reference cases were developed in FY2000 for the NETL/Gasification Technologies team and are documented on the website. They can accessed via the following URL.

http://www.netl.doe.gov/coalpower/gasification/system/destx3x_.pdf

As part of the DOE Clean Coal Technology demonstration projects, the Destec IGCC process was commercially demonstrated as the Wabash River Coal Gasification Repowering Project [9]. The DOE is currently sponsoring additional optimization studies [10] (Nexant, Global Energy) based on the results of this demonstration. This analysis and scope can accessed via the following URL.

http://www.netl.doe.gov/coalpower/gasification/projects/systems/docs/40342R01.PDF

For the present report the simulation codes developed earlier were updated to use version 10.2 of Aspen Plus® and the COE estimate was updated to first quarter 2002.

The cases have the following common process sections:
- Coal Slurry Prep - based on Illinois #6 coal, 66.6% solids.
- Destec Gasification - two stage, entrained flow, oxygen-blown, slagging gasifier.
- Air Separation Unit (ASU) - high pressure process integrated with the gas turbine.
- "G" gas turbine -W501G modified for coal derived fuel gas.
- Three pressure level subcritical reheat Steam Cycle
  - (1800 psia / 1050 °F / 342 psia / 1050 °F / 35 psia).

The approach used for gas cleanup accounts for the major differences between the two cases. For sulfur removal, Case 1 uses cold gas cleanup (CGCU) and Case 2 uses transport desulfurization hot gas cleanup (HGCU). The syngas gas cooler section following the gasifier (and integrated with the gasifier and other heat exchangers) is used for generating high-pressure superheated steam. This section is followed by a cyclone that captures particulates for recycle to the gasifier. The cooled raw fuel gas leaves the filter at a temperature of 650 °F for Case 1 and 1004 °F for Case 2. In Case 1, the raw fuel gas is further cooled (304 °F) and scrubbed and then sent to a gas cooling / heat recovery section before entering the CGCU section. In Case 2, the raw fuel gas enters a chloride guard bed prior to the HGCU section. Sulfur is recovered as elemental sulfur using the Claus process for Case 1 and as sulfuric acid using an acid plant for Case 2.

Process flow diagrams for these cases are shown in Figures 7 and 8. Additional flow diagrams (steam cycles) and material and energy balances summaries are provided in Appendix A and COE summaries are given in Appendix B. In Table 6 (above) the overall results obtained for power generation, process efficiency, and COE are compared for both cases.

17

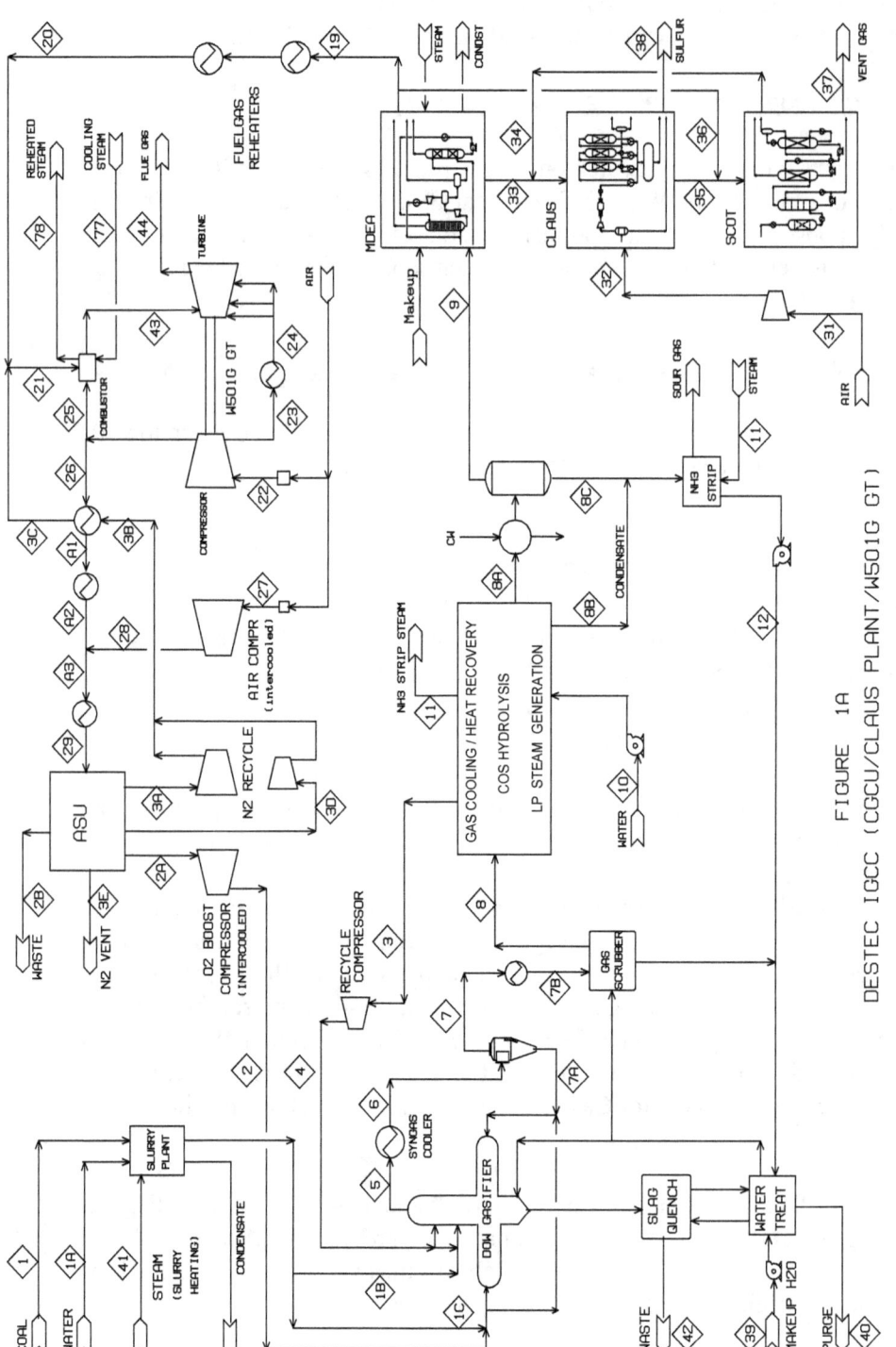

FIGURE 1A

DESTEC IGCC (CGCU/CLAUS PLANT/W501G GT)

Figure 7. Case 1. IGCC DESTEC / CGCU – No CO$_2$ Capture

18

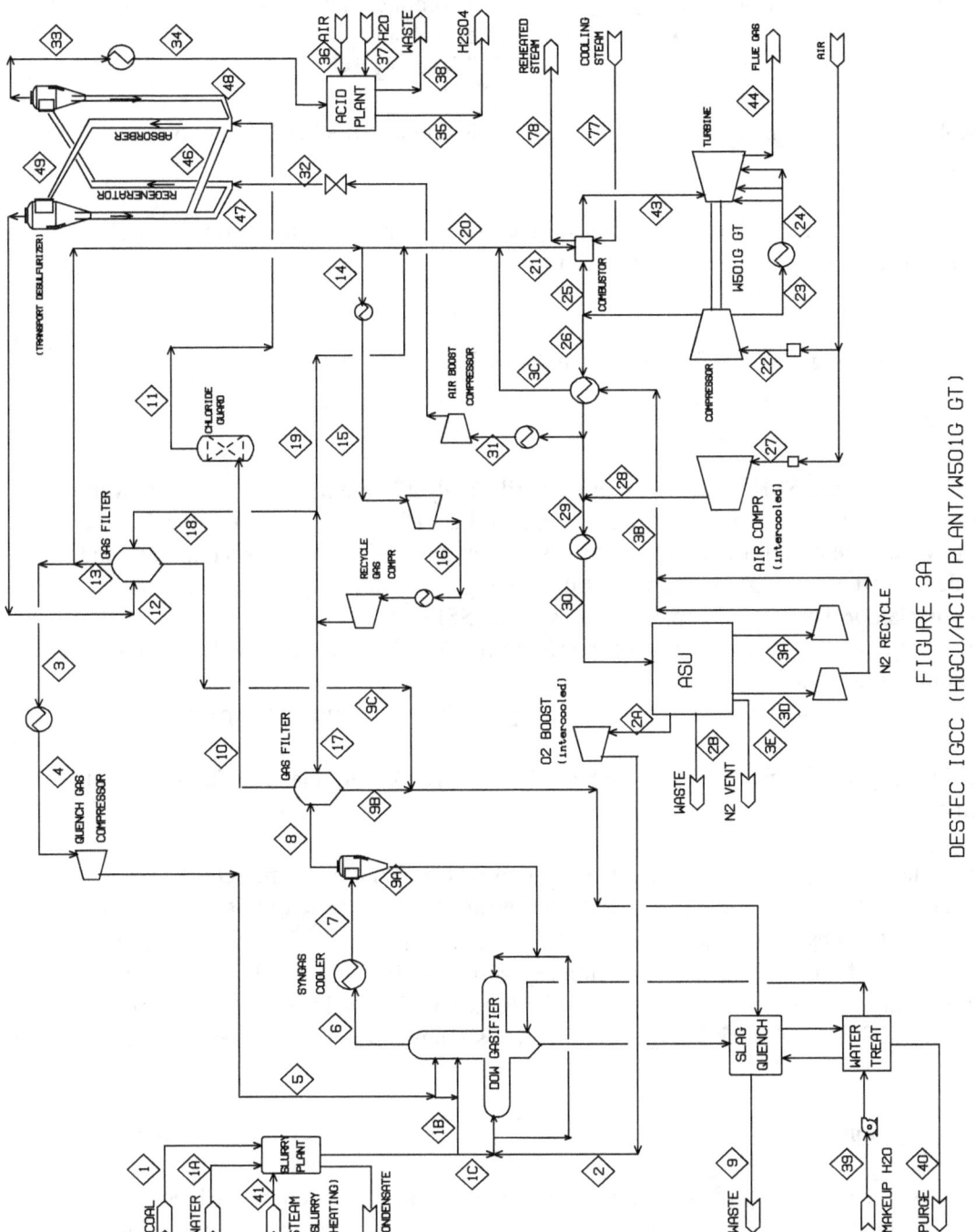

DESTEC IGCC (HGCU/ACID PLANT/W501G GT)

FIGURE 3A

19

Figure 8.   Case 2.   IGCC DESTEC / HGCU / ACID PLANT/ W501G GT – No CO$_2$ Capture

I-3.2 IGCC Destec (E-Gas™) Cases – CO_2 Capture

This case was developed based on modifying Case 1 to include CO_2 capture and involves the following changes in the power plant design:

- Shift Reaction Section using a catalytic process to accomplish the following reactions:
$$CO + H_2O \leftrightarrow CO_2 + H_2 \quad \text{(water-gas shift)}$$
$$COS + H_2O \leftrightarrow CO_2 + H_2S \quad \text{(COS Hydrolysis)}$$

- Selexol process for both H_2S and CO_2 removal. This replaces the MDEA section in Case 1.

- CO_2 compression in a multistage (5-stages) intercooled compressor to 2100 psia, cooling to 100 °F (liquid) and pumped to 3000 psia for storage.

- Gas Turbine – the gas turbine is fueled with the hydrogen rich fuel.

Shift Reaction Section

The catalyst chosen (named SSK , "Sulfur Tolerant Shift Catalyst") and process conditions were designed based on information provided to NETL (Patrick Le - 1997) by Haldor Topsoe, Inc. [11]. The catalyst can be used for both the water-gas shift and the COS hydrolysis reactions and was initially developed at EXXON Research & Engineering Laboratories and extended for industrial use by Haldor-Topsoe. The main features of the SSK catalyst are:
- unique property of being highly active for the reaction of carbon monoxide with steam in the presence of hydrogen sulfide.
- maintains its activity over a wide range of operating conditions including temperatures to 890 °F.
- No specific catalyst poisons are known for SSK. Insensitive to even relatively large amounts of chlorine.

The simulation model represents this section using a two-bed shift unit with intercoolers / aftercoolers for heat recovery that was integrated into the steam cycle. The required shift steam was bled from the steam cycle at conditions of 632 °F and 390 psia and mixed with the raw syngas and sent to the first catalytic bed. The first bed converts 70% of the CO and nearly all the COS. The exiting stream is cooled to 460 °F before entering the final stage. The overall conversion obtained for CO was 95%. After cooling, the stream is sent to the Selexol process section.

Selexol Process Section

This section is used to selectively remove H_2S in a product stream that is sent to a Claus unit for sulfur recovery and to recover CO_2 in a product stream that is sent to a compression unit for sequestration. The Selexol process is an absorber-stripper system that uses a designer physical solvent (Dow Chemical, formerly Union Carbide) containing a mixture of glycols. In the

Aspen Plus® simulation, the overall recoveries were represented and the detailed chemistry not modeled. The shifted cooled syngas is considered to enter an absorber that preferentially removes the H_2S by using a lean Selexol solvent that is loaded with CO_2. The rich solvent leaves the absorber and is sent to a stripper for regeneration. Low pressure steam used for the stripper reboiler is supplied from the steam cycle. The sweet syngas stream exits the first absorber and is sent to a second absorber that uses an unloaded solvent to remove CO_2 and additional H_2S. The CO_2 rich solvent stream leaves the second absorber and is recovered by flashing CO_2 vapor off the liquid at a reduced pressure. (Alternately, a second stripper could be used.) The cleaned syngas in the current simulation aimed at power production is reheated and sent to the gas turbine combustor. Alternately, if hydrogen is the desired product, the hydrogen rich syngas stream would be sent to a pressure swing absorption process for further purification with a residual fuel stream available for use in power generation. (see Case 5 that uses Shell gasification for this approach).

(It should be noted that the use of a double absorber system will result in improved H_2S removal which may approach the goals set for hot gas cleanup units {Case 2}. The sulfur emissions levels reported in Table 5 assumed that the SCOT waste stream was not recycled to the gasifier. Recycling would perhaps reduce the values shown by one-half. {HGCU levels}.)

CO_2 Compression Section

The CO_2 from the Selexol section is considered to be recovered in two streams from flashes at pressures of 40 psia (90%) and 15 psia (10%). The lower pressure stream is compressed to 45 psia and combined with the larger stream and sent to a multistage (5 stages) intercooled compressor to approximately 2100 psia. The supercritical stream is cooled to approximately 100 °F (liquid) and pumped to 3000 psia for storage. This section requires 19.9 MWe of power.

Gas Turbine Section

The gas turbine is fueled with the hydrogen rich syngas stream. To maintain approximately the same turbine power output and turbine inlet temperature as in Case 1 and Case 2, the coal flowrate (27% increase) to the gasifier and the nitrogen recycle from the ASU were adjusted.

This case results in an overall decrease in process efficiency (LHV) of 6.6 percentage points when compared with Case 1 (no CO_2 capture) which is attributable to the additional compression power requirements and the reduction in steam cycle output due to the steam requirements of the shift reaction section. The COE also shows a corresponding increase to 54.4 from 40.9 $/MW-hr.

Flow diagrams and M&E balance summaries are provided in Appendix A and the COE estimate is provided in Appendix B.

I-3.3 IGCC Shell Cases

Two reference cases are included based on the Shell Gasification process. Case 4 was developed in FY2000 (EG&G) [12] and Case 5 in FY2001 (ANL, J. Molburg, R. Doctor, N. Brockmeier) [13] for the NETL/Gasification Technologies team. The documentation can be accessed via the following URLs.

Case 4:
http://www.netl.doe.gov/coalpower/gasification/system/shell3x_.pdf

Case 5:
http://www.netl.doe.gov/coalpower/gasification/pubs/pdf/igcc-co2.pdf

Case 4 corresponds to an IGCC system that is analogous to Case 1 differing primarily in the use of a Shell gasifier replacing the Destec gasifier. Case 5 was developed using Case 4 as a starting point and making modifications to enable CO_2 capture making this case similar to Case 3 that used the Destec gasifier. Additionally, Case 5 has the objective of producing a hydrogen product stream of high purity as either a chemical product or as fuel for an advanced power module such as a fuel cell.

Case 4 (Shell IGCC) consists of the following major sections:

- Coal Prep - coal grinding and fluid-bed dryer to approximately 5% moisture.
- Shell Gasification - entrained flow, oxygen-blown, slagging gasifier.
- Air Separation Unit (ASU) - high pressure process integrated with the gas turbine.
- Cold Gas Cleanup – MDEA, Claus, SCOT – sulfur removal and recovery.
- "G" gas turbine -W501G modified for coal derived fuel gas.
- Three pressure level subcritical reheat Steam Cycle
    - (1800 psia/1050 °F/342 psia/1050 °F / 35 psia).

The raw fuel gas cooler section following the gasifier (and integrated with the gasifier and other heat exchangers) is used for generating high pressure superheated steam. This section is followed by a ceramic filter that captures particulates for recycle to the gasifier. The cooled raw fuel gas leaves the filter at a temperature of 640 °F. The raw fuel gas is further cooled, enters a COS hydrolyzer, and is scrubbed (removes remaining particulates, ammonia and chlorides) before entering the CGCU section. Sulfur is recovered as elemental sulfur using the Claus process for Case 1. The cleaned fuel gas is reheated and sent to the gas turbine for power generation. The turbine exhaust enters a HRSG that generates steam at three pressure levels for use in the steam cycle. The overall process efficiency is 47.4 % (LHV).

A process flow diagram for this case is shown in Figures 9. Additional flow diagrams (steam cycles) and material and energy balances summaries are provided in Appendix A and a COE summary is in Appendix B. In Table 6 (above) the overall results obtained for power generation, process efficiency, and COE are listed.

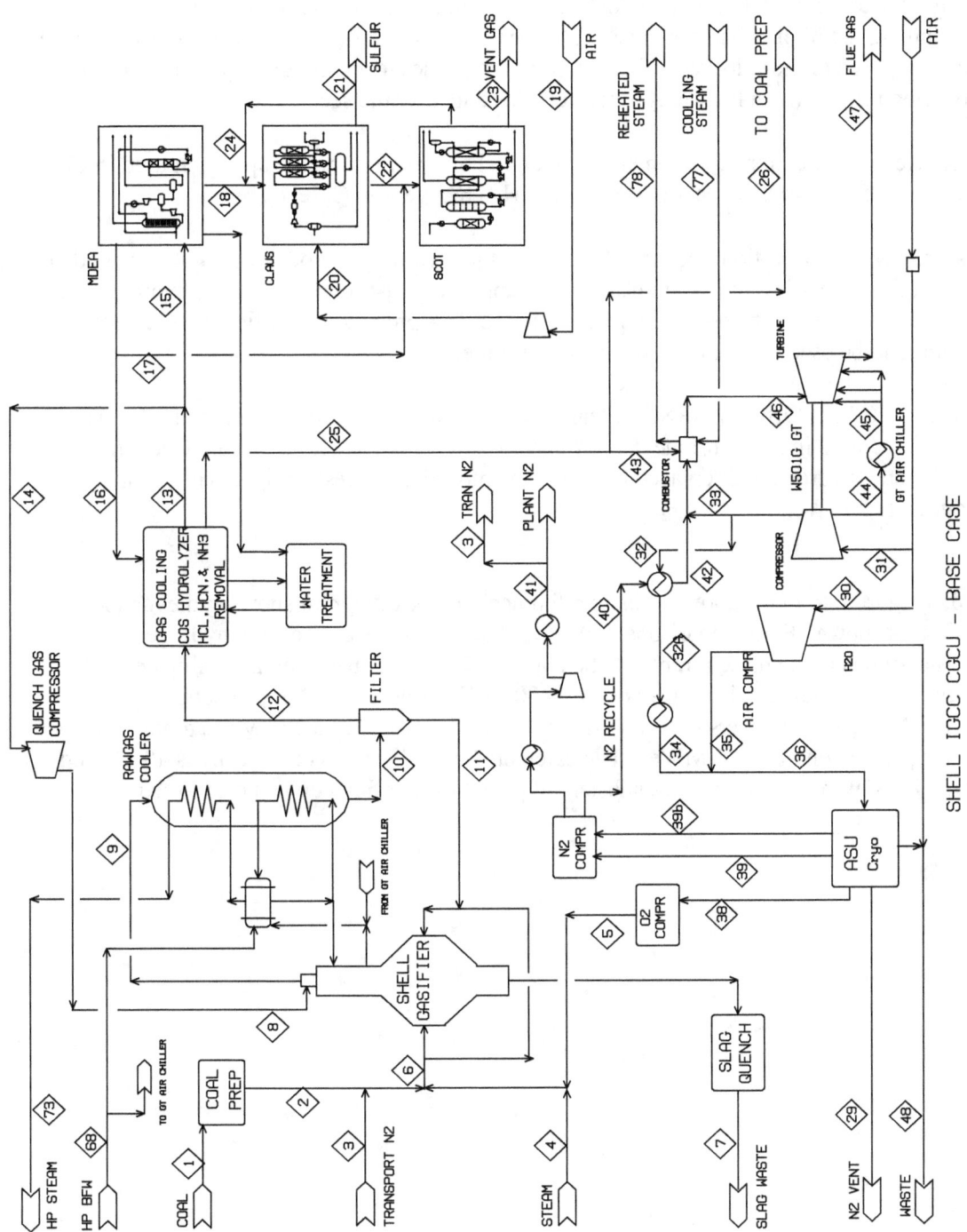

Figure 9. Case 4. IGCC SHELL / CGCU – No CO$_2$ Capture

23

For Case 5, ANL made the following modifications to Case 4:

- Shift Reaction Section - The shift reaction is used to convert CO in the gasifier product stream to CO_2 and hydrogen using two beds of sulfur-tolerant shift catalyst. The first bed was used to convert 76% of the CO and 98% of the remaining CO in the second bed. Steam requirements are higher than for Case 3 (Destec) since the gasifier in this case uses a dry coal feed as opposed to the slurry coal feed. Again part of the steam energy requirement is met by recovering heat between the catalyst bed sections and after the second bed.

- Glycol Recovery Sections for both H_2 and CO_2 - This is similar to the approach used in Case 3 and replaces the MDEA section used for the H_2S recovery in Case 4.

- Pressure Swing Absorption Section – Since the objective was to produce a highly purified H_2 stream, this process is required. In Case 3, this approach wasn't used since the hydrogen was used in a gas turbine. The residual stream from the PSA process has sufficient heating value remaining to be used as fuel in a midsize gas turbine.

- Replacing "G" gas turbine / HRSG / Steam Cycle – The residual fuel from the PSA was reheated and used in a gas turbine that produces 62 MWe . The HRSG/Steam Cycle from Case 4 were discarded and replaced to reflect the modified process design. The steam cycle produces 91.5 MWe.

In Figure 10, (Figure 1 from the above website reference), a block diagram showing the major process sections is shown. For comparisons with other IGCC reference cases, the hydrogen produced was assumed in the present report to be converted to power based on assuming an advanced process (e.g., fuel cell) having a cost of $400/MWe. Based on ANL projections, (see Table 2 of the ANL report), conversion at an efficiency of 65% would add 275 MWe to the process for a net power production of 351.1 MWe . The calculated overall process efficiency is 40.1% and the COE is 62.9 $/MW-hr. This indicates substantial penalties in efficiency and cost to sequester the CO_2.

Shell-based Gasification Combined Cycle Plant with Hydrogen and Carbon Dioxide

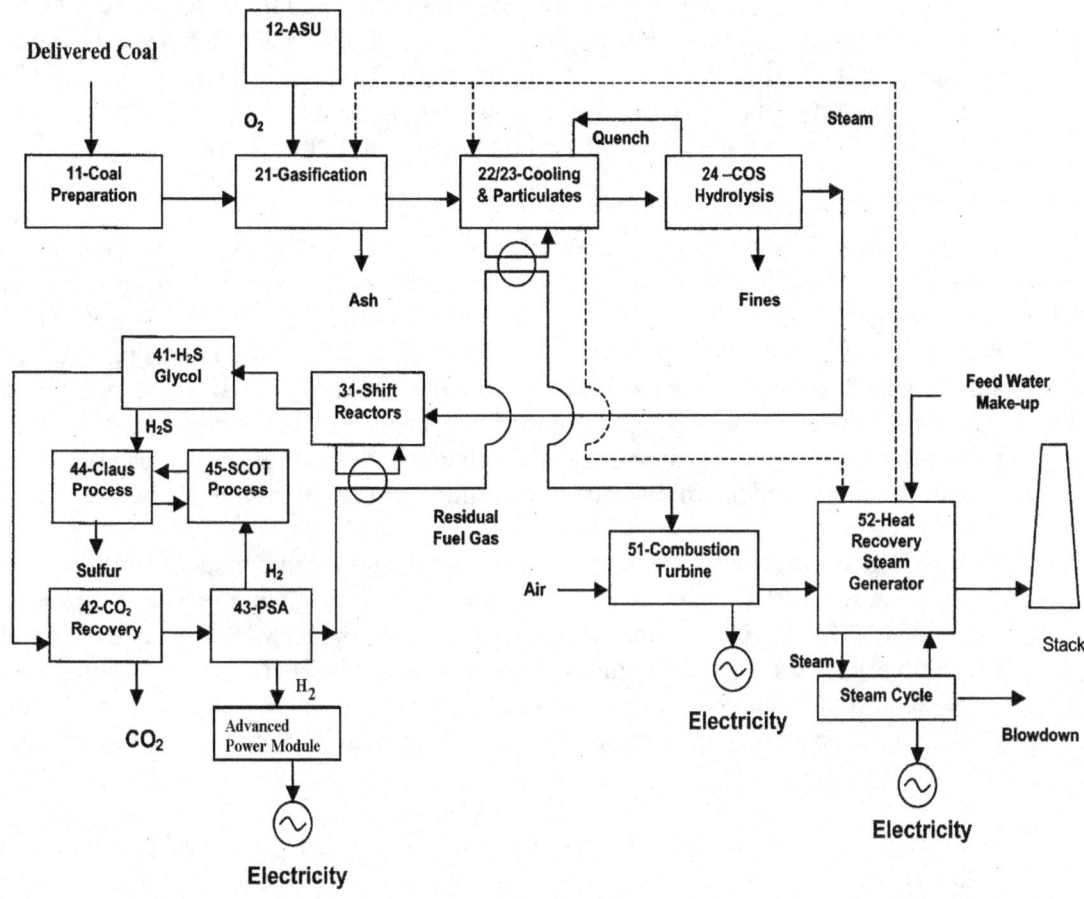

Figure 10.   Case 5.   SHELL / CO₂ Capture / Advanced Power Module

I-4. Summary – Reference Plants

The reference plants included in the previous sections were provided to have points for comparison for the advanced fossil power systems considered in the remainder of this report. The systems were projected for a nominal plant size of 400 MWe (for cases having no carbon dioxide capture) and with a consistent cost of electricity analysis based on the EPRI TAG method (see Appendix B). Additionally, cases were included to illustrate the significant penalty that occurs with the addition of carbon dioxide sequestration that may be required for Vision 21 power plants.

The PC power plant (no CO_2 capture) represents a primary system presently employed for coal based power plants in this country. It is expected that these plants will be subjected to further requirements for improved emissions than the results shown in Table 2. The efficiency determined of 39% (LHV) can be improved to about 43-47 % based on using a super-critical steam cycle, higher steam temperatures and double reheat cycles. All these involve additional costs. The two remaining PC cases included CO_2 capture either using flue gas cleanup or a proposed system based on using oxygen. Both cases illustrate an energy penalty of 8 – 10 percentage points and approximately double the COE results from the base system.

Two NGCC systems were included based on using a gas turbine model of the Siemens-Westinghouse W501 G gas turbine. The inclusion of CO_2 capture reduces the process efficiency from 58% (LHV) to 50% and increases the COE from 33.1 ($/MW-hr, constant $) to 46.4. Projections provided by both Siemens-Westinghouse and General Electric to the DOE anticipated commercial NGCC systems (no CO_2 capture) with efficiency above 60% (LHV). NETL/DOE is currently sponsoring research [4] aimed at improving the flue gas CO_2 capture to reduce the energy penalty.

The IGCC cases included were for systems aimed at providing electrical power and not a mix of both power and chemicals. The penalty (for the cases considered) associated with CO_2 capture is 6.5 – 7.3 percentage points. Since the CO_2 capture involves treating the generated fuel gas rather than the flue gas of a NGCC process, the capture is easier and more feasible both form a technical and economic viewpoint. However, this is balanced by the inherent difference in the carbon/hydrogen content of coal versus natural gas. The arguments made for IGCC systems are usually made based on the potential offered for feedstock diversity (and product diversity) and the energy security based on using our (USA) most abundant resource, coal. The economic comparison with the NGCC is dependent on the price assumed for natural gas. (A value of $3.2/MM BTU was used for natural gas cases.). Using the near-term commercial systems for IGCC, the expected efficiency is significantly lower than the 60% (HHV) goal of Vision 21 plants based on coal.

II. ADVANCED POWER CYCLES

II-1 Hydraulic Air Compression Cycle (HAC)

The use of hydraulic air compression (HAC) has been proposed as a means for increasing the efficiency of high-efficiency power cycles to meet the Vision 21 objectives for both natural gas and coal [14]. In this approach, low pressure air is entrained in a large volume of water with the resulting mixture pressurized using a deep well or reservoir. The high pressure air produced can be used to replace the high pressure air normally supplied by the gas turbine compressor in a combined cycle power system. Conceptually, the gas turbine in either the NGCC or IGCC is modified by removing the compressor while retaining the combustor and expander sections. Additionally, the proposed HAC power cycles employ the expander exhaust in a recuperator to preheat the high pressure air sent to the combustor. This either eliminates the need for a steam cycle or greatly reduces its size and cost. A simplified diagram illustrating the HAC is shown in Figure 11.

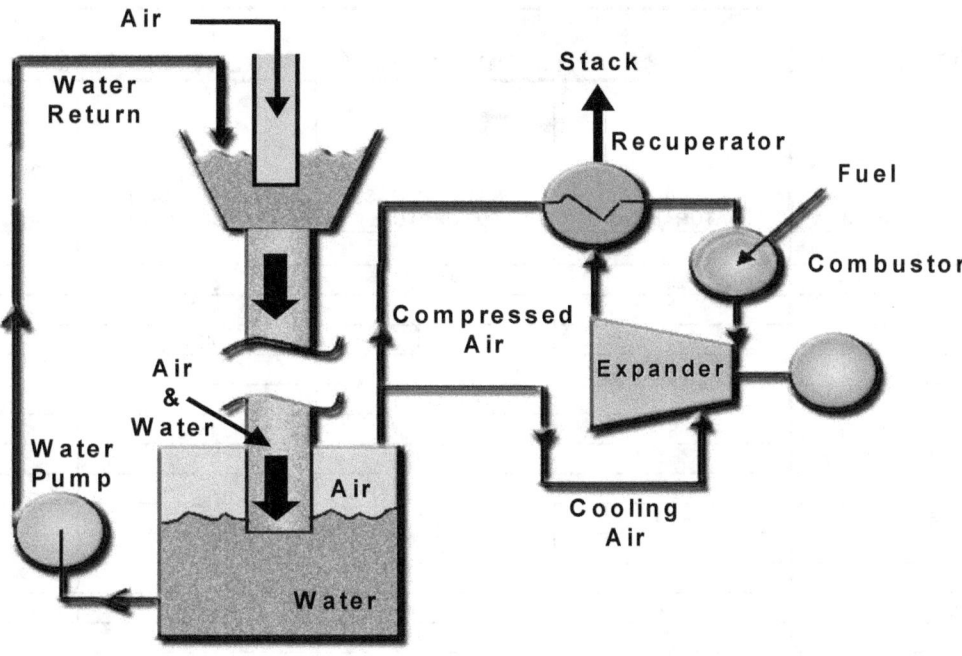

Figure 11. Hydraulic Air Compression Power Block – closed loop water cycle.

The following simulation cases were developed to provide high pressure air to the combustor using the Hydraulic Air Compression:

- Case 1 - Natural Gas Cycle without CO_2 capture. This case modifies the NGCC reference plant case.
- Case 2 - Natural Gas Cycle with CO_2 capture. This case extends Case 1 by adding an amine plant / compression sections to recovery the CO_2.
- Case 3 - Coal Cycle without CO_2 capture. This case modifies the Destec IGCC (CGCU) reference plant case.
- Case 4 - Coal Cycle with CO_2 capture. This case modifies a Destec IGCC (High Pressure Gasifier/Gas Shift Reaction/HGCU) process plant. This is a case developed for this report.

The results obtained from these simulations are provided in Table 7.

Table 6. Hydraulic Air Compression Cycles

POWER SYSTEM	HYDRAULIC AIR COMPRESSION (HAC)			
Generation Cycle	HAC NATURAL GAS	HAC NATURAL GAS (CO2 CAPTURE)	HAC Destec (E-Gas) CGCU	HAC Destec HP (E-Gas) HGCU (CO2 CAPTURE)
Case	1	2	3	4
Net Power MWe	323.5	300.2	325.9	312.4
Net Plant Efficiency	53.2	43.8	43.8	35.2
% LHV				
Total Capital Requirement	681	1140	1436	2189
$ / KW				
Cost of Electricity	44.2	61.0	47.0	65.5
$ / MW-hr				
NOx emissions	0.194	0.210	0.193	0.204
lb/MW-hr				
Sox emissions	---	---	0.337	0.048
lb/MW-hr				
CO2 Production				
lb/MW-hr				
a) Emitted to atmosphere	824	100	1561	142
b) Sequesterable		899		1870
Footprint (battery limits)	179	230	1293	1583
sq ft/MW				

II-1.1 Hydraulic Air Compression Cycle (HAC) – Natural Gas

Aspen Plus® simulations were developed to estimate the approximate performance and cost estimate for cases with and without CO_2 capture. These cases essentially modify the reference NGCC cases by replacing the air compressor with air obtained from the HAC approach. The combustor and turbine sections were assumed to be the same as the W501 G gas turbine.

The HAC process assumed that the air normally required for the W501 G air compressor was blown into an air/water induction system. The water usage into the closed loop system was set using the estimation method provided in a NETL sponsored study [14]. The resulting water/air mass ratio obtained was 1115. [15]. This large water usage leads to a requirement for a number of large pumps for recirculation. The high pressure air produced and delivered to the combustor was preheated in a recuperator using the exhaust stream from the gas turbine expander. For the case without CO_2 capture, the air is preheated to 950 °F and the cooled exhaust stream enters a small heat recovery section to generate low pressure (35 psia) steam used for combustor duct cooling. After being heated in the combustor duct, the steam is sent to a small steam turbine. For the case with CO_2 capture, the air was only preheated to 725 °F and a larger HRSG used since a large amount of steam is required for the stripper reboiler in the amine based CO_2 recovery process (see Figure 6 – amine plant).

Emissions of CO_2 were based on simulation results and NOx was estimated as 9 ppmv as projected for "G" turbine combustor performance. The cost estimates were based on modifying the NGCC reference plant cases. Reductions were subtracted from the total capital for the elimination of the air compressor, HRSG and steam turbines. Additions for the following: hydraulic air compression blowers and pumps (40 MWe), recuperators (large area heat exchangers), reservoir well (650 ft depth, 20 ft diameter), and for miscellaneous HAC equipment ($50 / KW). The footprint estimates were assumed to be equal approximately to those of the NGCC reference plants with an additional 1 acre for the HAC related equipment. Again the total plant sites were assumed to cover 100 acres.

The overall process efficiencies (LHV) obtained were 53.2 % (no CO_2 recovery) and 43.8 % (with CO_2 recovery). The total capital requirements and COE estimates made with conservative assumptions are provided in Table 7. The results for both efficiency and COE are higher than comparable reference cases given in Table 1. The lower efficiency is related to the large power requirements of the recirculation water pumps and the requirement to add a recuperator to preheat the high pressure air. The inclusion of the recuperator using the turbine exhaust essentially eliminated the power produced by the steam turbines in the reference cases. These closed loop HAC systems will be unable to obtain the goals of the Vision 21 power plants.

The two cases are shown in Figure 12 and Figure 13. Appendix A contains material and energy flow rate summaries and Appendix B includes the COE spreadsheet summaries.

CASE 1

HYDRAULIC AIR COMPRESSION CYCLE - NATURAL GAS - NO CO2 SEQUESTRATION

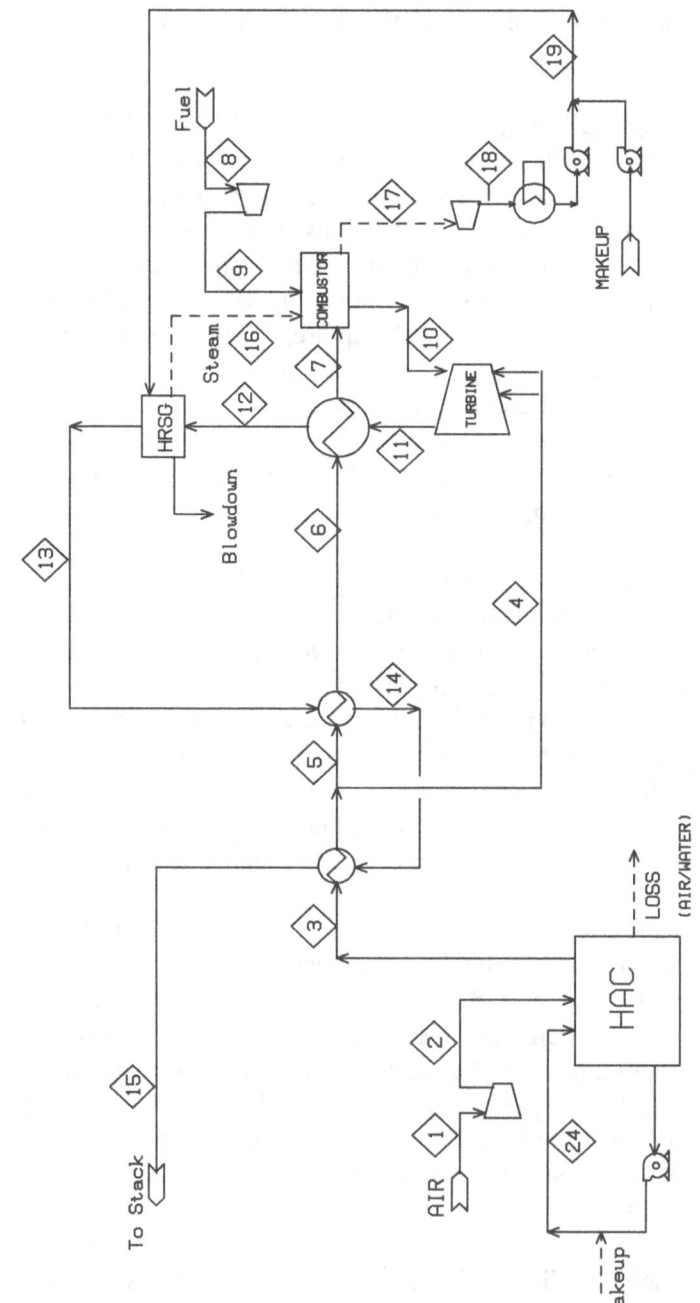

Figure 12. Case 1 - Natural Gas HAC – without CO₂ Capture

CASE 2

HYDRAULIC AIR COMPRESSION CYCLE - NATURAL GAS - CO2 SEQUESTRATION

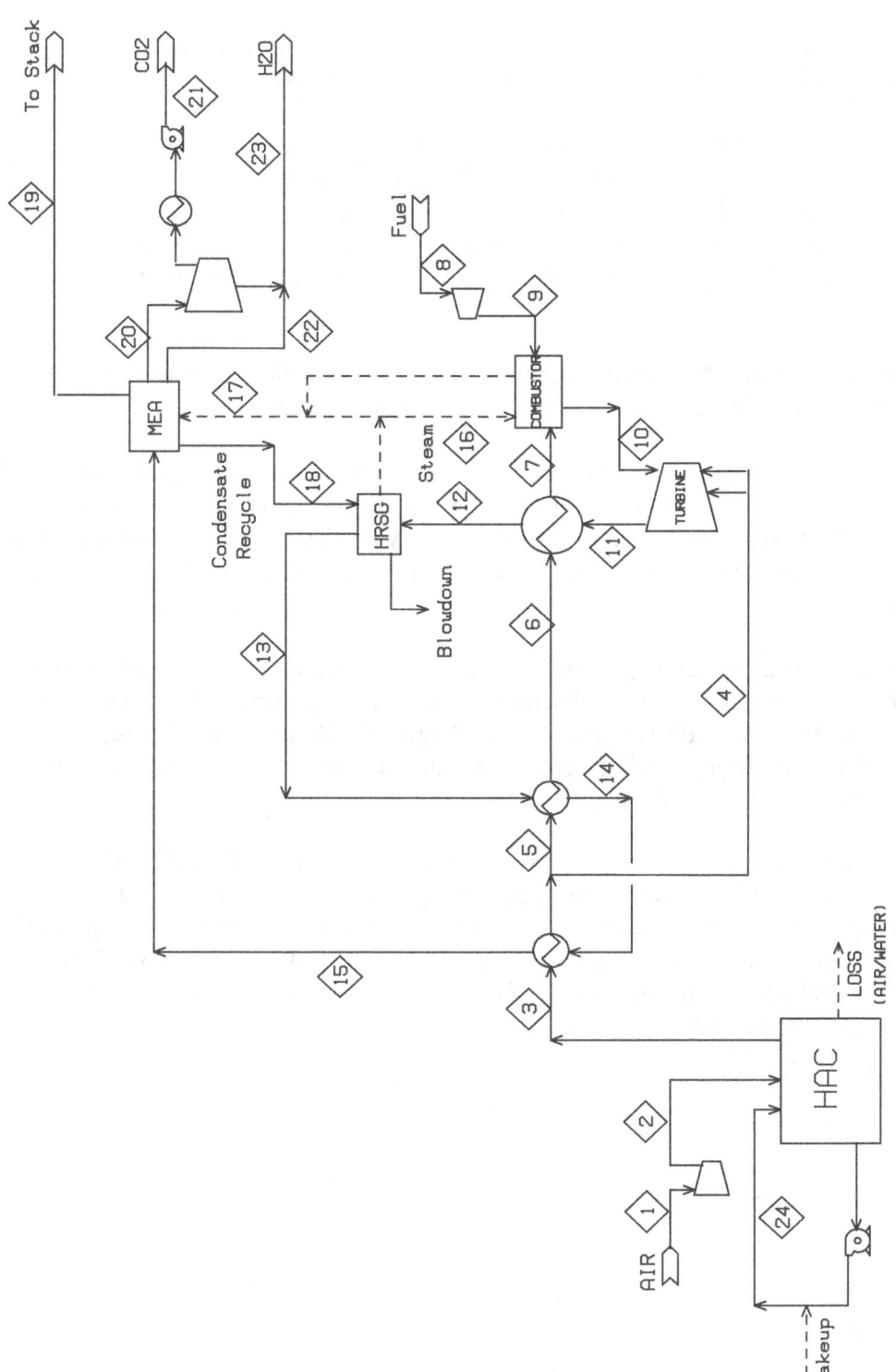

Figure 13. Case 2 – Natural Gas HAC – with CO$_2$ Capture

II-1.2 Hydraulic Air Compression Cycle (HAC) – Coal – without CO₂ Capture

This case is based on modifying the IGCC reference case based on the Destec gasification process that uses CGCU for sulfur recovery. The modifications include:

- The HAC is used to replace the gas turbine's air compressor. High pressure air is supplied to both the gas turbine combustor and the air separation unit (ASU). As in the natural gas cases, the air flowrate required for the combustor and ASU is fed to the HAC module. Nitrogen available from the ASU was used to replace chargeable cooling air for cooling in the turbine expander. The water flow rate is set at 1115 times the air flowrate. (mass basis).

- A recuperator is added that uses the turbine exhaust to preheat air sent to the combustor. The turbine exhaust leaves the recuperator at 265 °F and is sent to a stack.

- The reference case steam cycle (HRSG/steam turbines) that generates steam at three pressure levels is replaced with a smaller system (33 MWe) based on generating steam at a single high pressure. The steam generation is mainly now due to the syngas cooler since the heat available in the turbine exhaust was used in the recuperator section for preheating air.

- The cost estimate is based on adjusting the reference case for sections removed and used the same algorithms for HAC related items as in the natural gas case. The footprint was somewhat smaller due to the elimination of the larger HRSG/Steam Turbine sections found in the reference case. Additionally, since the net power increased, the footprint on a (ft² / MWe) basis is approximately 20% smaller.

The net power produced decreased from the reference IGCC case by 77 MWe and the COE increased to 47.0 from 40.9 ($/MW-hr). The overall process efficiency obtained was 43.8 % (LHV) or 42.3% (HHV). Again the efficiency falls significantly below the 60% (HHV) goal of Vision 21 for a power system based on coal. In Figure 14 and Figure 15, process flow diagrams are shown. In Appendix A, summaries are provided for material and energy flowrates. In Appendix B, the COE spreadsheet is provided.

Case 3

HYDRAULIC AIR COMPRESSION CYCLE - COAL SYNGAS - NO CO₂ SEQUESTRATION

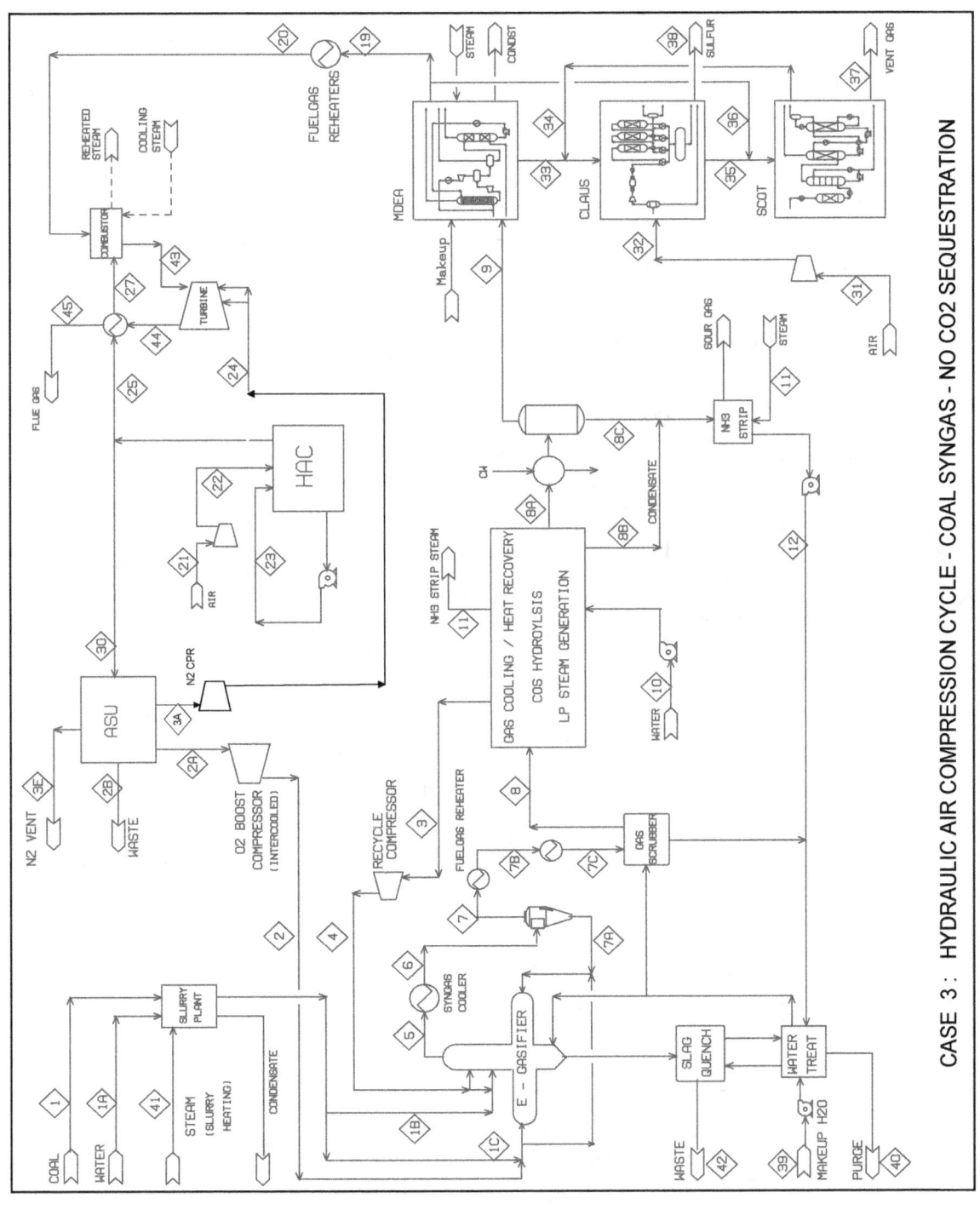

CASE 3 : HYDRAULIC AIR COMPRESSION CYCLE - COAL SYNGAS - NO CO2 SEQUESTRATION

Figure 14. Case 3 - Coal Syngas HAC – without CO₂ Capture

33

Case 3

HYDRAULIC AIR COMPRESSION CYCLE - COAL SYNGAS - NO CO₂ SEQUESTRATION
STEAM CYCLE

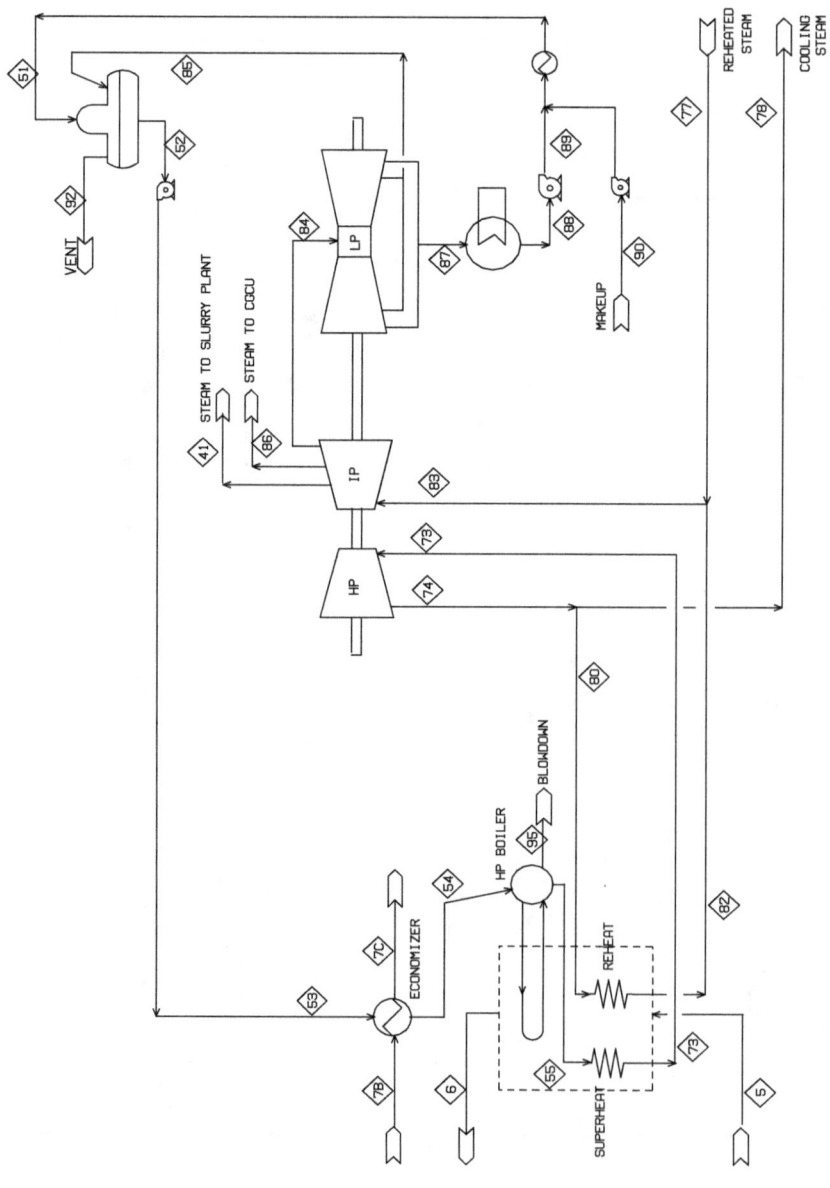

Figure 15. Case 3 - Steam Cycle

II-1. 3 Hydraulic Air Compression Cycle (HAC) – Coal – with CO_2 Capture

The reference Destec IGCC cases showed an advantaged of 2.7 percentage points in overall process efficiency when using HGCU in place of CGCU for sulfur removal and lower SOx emission levels. (see Table 6). This was the primary reason for using the Destec IGCC reference case based on HGCU as the starting point for developing the present HAC case since a significant energy penalty is expected for sequestrating the CO_2. An additional reason was that having a cleaned coal syngas at high temperature would allow the use of a Hydrogen Separation Device (HSD) currently being developed with DOE funding at ORNL [16]. The HSD is a membrane catalytic reactor being designed to both shift the coal syngas and separate out a high purity hydrogen stream. The modifications made to the reference case include the following:

- Gasifier pressure was increased to enable the downstream HSD device to have an inlet pressure of approximately 1000 psia. This also increases the power requirements for the oxygen boost compressor that supplies the gasifier. The cost analysis considers that two gasifier trains will be required based on information provided by Destec (now Global) to the DOE in previous contractor studies [17].

- A model for the HSD was added following the HGCU section. Steam at 1000 psia was added for accomplishing the shifting of the coal syngas stream. The HSD produces two streams, a high pressure CO_2 rich-stream and a low pressure high purity H_2 rich-stream.

- The CO_2-rich stream (with residual fuel gas) is sent to a power turbine and proceeds to an oxygen fired combustor to burn any residual fuel before entering a HRSG for steam generation. This stream is further cooled before entering a multi-stage compression section that raises the pressure to 2100 psia. Subsequent cooling to 100 $^{\circ}$F produces a liquefied product stream.

- The hydrogen-rich stream is sent to a separate HRSG for steam generation before entering a compression section. The hydrogen is now available for use as a fuel in the HAC module.

- The HAC module is based on Case 3 (see above).

- The steam cycle developed recovers energy from the gasifier syngas cooler, the acid plant section, and the two HRSG sections that follow the HSD device.

The inclusion of the HAC system again results in a power plant having a significant loss in net power due. This case produced 312 MWe at an overall efficiency of 35.2 % (LHV). Compared to Case 3, the CO_2 recovery resulted in an energy penalty of 8.6 percentage points and an increase in the COE estimate to 65.5 from 47.0 ($/MW-hr). Process flow diagrams are shown in Figure 16 and Figure 17. Appendix A lists summaries for the material and energy flowrates and Appendix B lists the COE spreadsheet results.

Case 4

HYDRAULIC AIR COMPRESSION CYCLE - COAL SYNGAS - CO₂ SEQUESTRATION

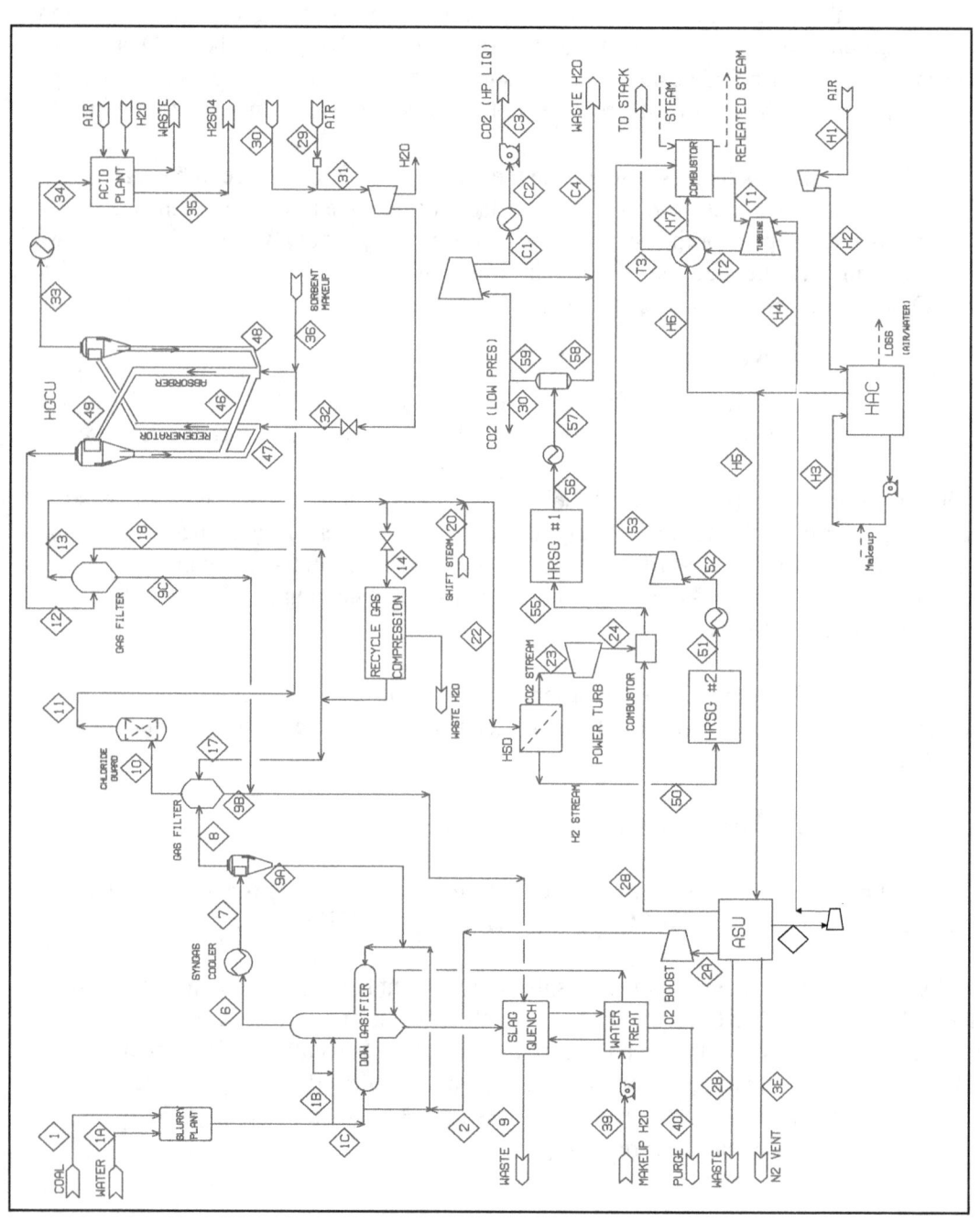

Figure 16. Case 4 – Coal Syngas HAC – with CO₂ Capture

36

Case 4

HYDRAULIC AIR COMPRESSION CYCLE - COAL SYNGAS - CO₂ SEQUESTRATION

HRSG/STEAM CYCLE

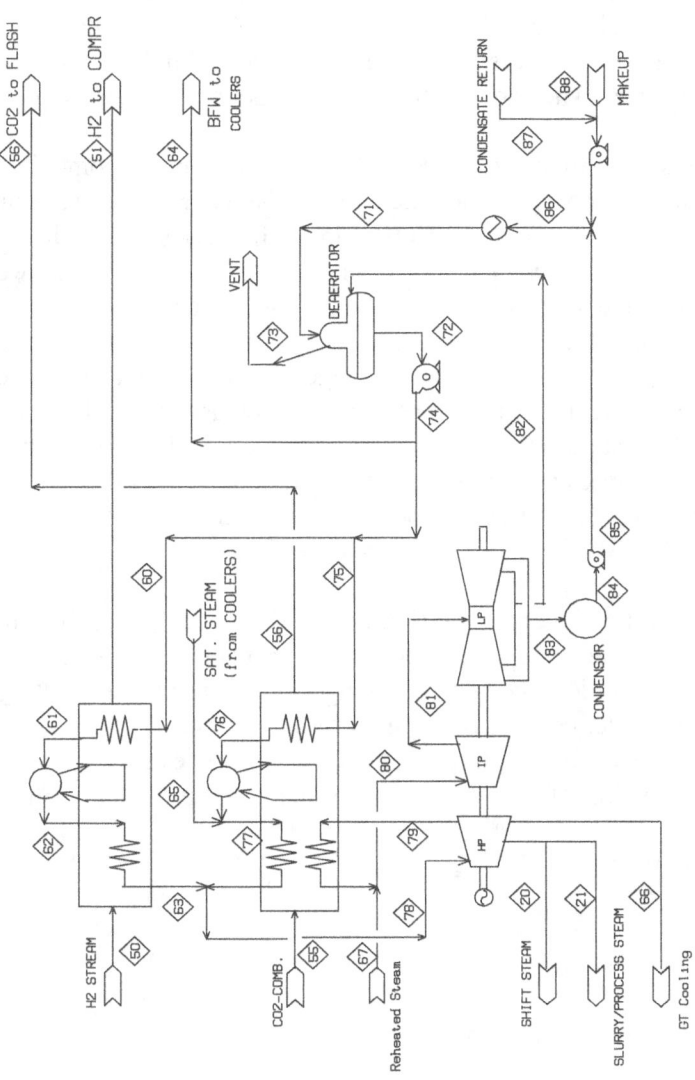

Figure 17. Case 4 - Steam Cycle

II-1.4 Hydraulic Air Compression Cycle (HAC) – Summary

In Table 8, the simulation cases are summarized with the performance and the power listed for major process areas. The overall process efficiencies obtained for all cases do not approach the goals of the Vision 21 program and are lower when compared with reference cases.

The use of the HAC module requires from 170 – 202 MW due primarily for water pumps and varies with the case's air requirement. The air required for the coal cases is higher since the HAC supplies both the gas turbine combustor and the ASU. The HAC power requirements are somewhat less than the original air compressor (> 240 MW) that has been assumed to be removed from the gas turbine. For all cases a recuperator preheats the high pressure air with the turbine exhaust as part of the HAC module resulting in the loss or major reduction of the power generated from steam turbines normally found in the NGCC or IGCC power plants. This offsets the power gained by removing the air compressor. The results in Table 8 indicate net power losses of approximately 30 – 90 MWe when compared with corresponding reference plants.

Inclusion of CO_2 capture lowers the efficiency significantly by 9.4 percentage points for natural gas and by 8.7 percentage points for coal. The large penalty for the natural gas case is directly related to the poor performance inherit in removing CO_2 from the flue gas stream. The compression power (compression to 2100 psia) and the amine power (inlet flue gas blower, included in MISC/AUX in Table 8) requirements significantly reduce the net power generated. Removing CO_2 in the coal case was based on treating the coal syngas by a membrane reactor system (an advanced technology presently in the research stage of development) that produces a H_2 rich fuel stream and a CO_2 rich stream. This case required an increase in coal flowrate compared to the case without CO_2 capture to obtain sufficient fuel to obtain the same turbine expansion power. Additionally, more CO_2 is produced using the coal fueled process compared to the natural gas fueled process. This is reflected in power requirements for the CO_2 compression section and the MISC/AUX section shown in Table 7.

The cost analysis included process contingencies of 25% for the HAC section and 50% for the HSD section to reflect that these two areas represent technology that is in a development stage and not commercially available. Additionally, a 25% contingency was used in estimating the costs for the modified turbine expander/combustor required for these cases. The water pumps costs are also very significant and were based on using the ICARUS cost estimating package and on information obtained from a vendor [18]. The COE spreadsheets are provided in Appendix B.

Table 7. Summary of HAC Cases - with/without CO_2 Sequestration

CASE	1	2	3	4
FUEL	CH4	CH4	COAL	COAL
CO_2 CAPTURE	NO	YES	NO	YES
HHV %	48.1	39.6	42.3	33.9
LHV %	53.2	43.8	43.8	35.2
NET POWER MWe	323.5	300.2	325.9	312.4
work/power MWe:				
Gas Turbine Exp	494.8	498.8	499.1	501.7
CO_2 Expander	-	-	-	58.5
Steam Turbines	6.1	-	30.9	47.6
HAC	170.7	170.7	184.1	204.1
CO_2 Separation	-	11.4	-	28.2
H_2 Compression	-	-	-	26.1
MISC / AUX	6.6	16.5	20	36.9

II-1.5 Hydraulic Air Compression Cycle (HAC) – Open Loop Water System

In the study sponsored by NETL [14], HAC was considered for open loop water systems that could be located at dams or reservoirs. This eliminates a major power requirement for pumps found in the cases considered above based on a closed loop water system. An example from the study shows the following conceptual representation of this HAC module:

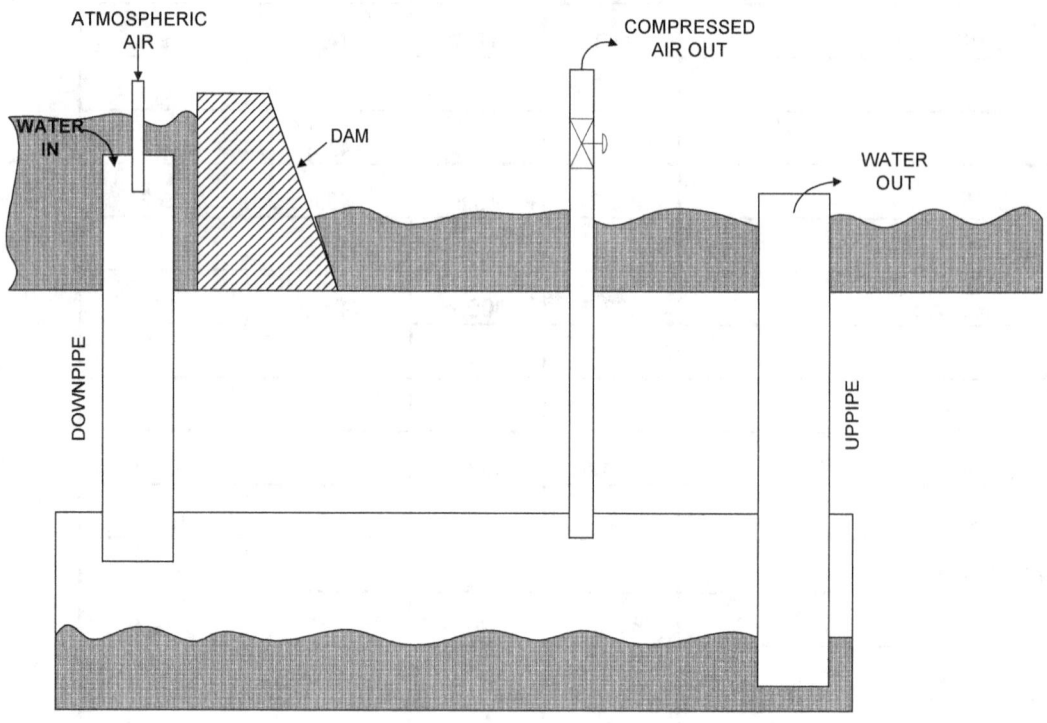

In Table 8, the results that were obtained for the closed loop water HAC cases have been modified to approximately judge what the results would have been for an open loop water system that could exist for a niche market at a dam site. The modifications made were to eliminate the HAC power requirements and obtain an adjusted net power and efficiency. These results were modified further by reducing the net power by the amount of power that would be expected to be generated using the same amount of water in a hydroelectric plant. The results show efficiencies that are about 10 – 13 percentage points (LHV) above the results obtained for the closed loop water systems. Additionally, these modified cases have higher efficiencies when compared to the reference cases by 5 – 8 percentage points. This indicates that the HAC approach for open loop water systems may be advantageous even though it will be a small market due to limited availability of applicable sites.

Table 8. Summary of HAC Cases - modified for open loop water system

CASE	1	2	3	4
FUEL	CH4	CH4	COAL	COAL
CO$_2$ CAPTURE	NO	YES	NO	YES
Power Adjustments (MWe) - for open (no water return) HAC				
gross power	330.1	306.3	335.9	322.1
hac cpr	2.1	2.1	2.2	2.5
hac pump	168.6	168.6	182.0	201.6
adjusted gross power	498.7	474.9	517.9	523.7
adjusted aux	10.0	9.5	15.5	15.7
adjusted net power	488.8	465.4	502.4	508.0
Adjusted Efficiency (hydroelectric power reduction not included)				
- HHV %	72.6	61.4	65.2	55.1
- LHV %	80.4	68.0	67.6	57.2
Calculation of Hydroelectric Power (same water usage & head as HAC)				
HAC Water Usage (M3/sec)	591.4	591.4	638.3	707.2
Hydraulic Head (M)	25.0	25.0	25.0	25.0
Water Power (MWe)	145.0	145.0	156.5	173.4
Hyroelectric Power (MWe)	87.2	87.2	94.2	104.3
http://www.iclei.org/efacts/hydro ele.htm		POWER (kW) = 5.9 x FLOW x HEAD		
		(60% of water power)		
Adjusted Net Power (includes hydroelectric reduction)				
- MWe	401.5	378.2	408.2	403.7
Adjusted Efficiency (includes hydroelectric reduction)				
- HHV %	59.7	49.9	53.0	43.8
- LHV %	66.0	55.2	54.9	45.4
Adjusted Total Capital Requirement $/KW	273.0	612.1	881.4	1449.6
Adjusted COE $ / MW-hr	25.8	38.0	28.5	41.8
Efficiency - non HAC system reference cases				
- LHV %	57.9	49.9	46.7	40.1
delta (HAC and Non-HAC)	8.1	5.3	8.2	5.3

II-2. CLEAN ENERGY SYSTEMS (CES) – ROCKET ENGINE STEAM CYCLE

Clean Energy Systems (CES) [19] has proposed an electric power generation system based on using fossil fuels such as natural gas, coal syngas (cleaned of sulfur), and coal-bed methane. The system, termed Zero Emission Steam Technology (ZEST) uses a combustion process that burns nearly pure oxygen with a hydrocarbon fuel under stoichiometric conditions. This essentially eliminates the formation of oxides of nitrogen and produces a product that contains primarily carbon dioxide and steam. In the CES process, Figure 18, a gas generator injected with a recycled high pressure water/steam mixture is fired with a fossil fuel using high pressure oxygen. The exhaust powers a high pressure/high temperature turbine (HPT). The HPT exit stream is used for water/steam heating and sent to a combustor reheater to increase the temperature to levels expected for advanced combustion turbines (i.e. >2500 $^\circ$F). The remaining turbine sections may have intermediate feed water heaters before the exhaust stream (approximately 90% H_2O, 10% CO_2) enters a partial condenser and then a condenser / CO_2 recovery section.

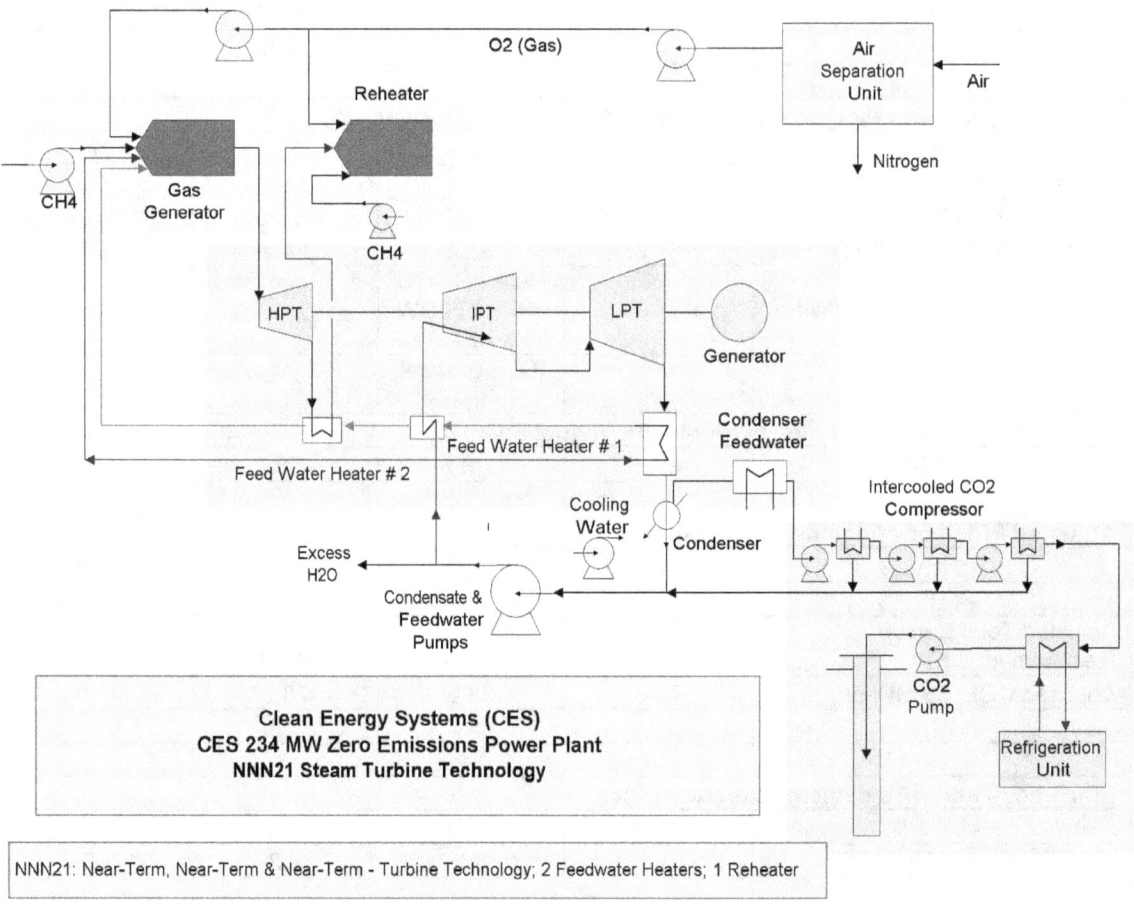

Figure 18. CES Process (provided by CES – version NNN21).

Current overall efficiency projections (LHV basis) provided by CES to NETL [20] for natural gas systems ranged from 44 % - 62 % and recently published results for coal systems [21] ranged from 32% to 44%. The higher values assume turbine technology developments that allow for inlet temperatures of 3200 °F, low last turbine stage exhausts (0.65 psia) and the use of oxygen generation using membranes.

Aspen Plus[®] simulations were developed based on flow diagrams provided by CES (Larry Hoffman, CES) for both a natural gas system and a coal system. Emissions for NOx were considered negligible since high purity oxygen (99.5%) was used in the simulations. CO_2 was estimated from the ASPEN simulations and considered sequestered as a liquid using a CO_2 compression scheme.

The COE estimates were developed using information provided by CES in reports and communications to NETL. [22]. Footprint (battery limits) were developed for the natural gas case based on the ASU plant being the major equipment section. The coal case used this approach and the footprint determined for a Destec IGCC plant.

In Table 9, results obtained are listed:

Table 9. CES – Rocket Engine Systems

POWER SYSTEM	ROCKET ENGINE (CES)	
Generation Cycle	CES Natural Gas (gas generator) (CO2 CAPTURE)	CES / COAL (gas generator) Destec HP (E-Gas) HGCU (CO2 CAPTURE)
Net Power MWe	398.4	406.2
Net Plant Efficiency % LHV	48.27	41.4
Total Capital Requirement $ / KW	975	1768
Cost of Electricity Constant $ / MW-hr	49.2	49.3
NOx emissions lb/MW-hr	NEG	NEG
Sox emissions lb/MW-hr	---	0.044
CO2 Production lb/MW-hr		
a) Emitted to atmosphere	---	---
b) Sequesterable	901	1702
Footprint (battery limits) sq ft/MW	825	1458

II-2.1 Clean Energy Systems (CES) - Natural Gas System

An Aspen Plus® simulation was developed for the natural gas fueled CES proposed process as shown in Figure 19. The key process sections are:

- Cryogenic ASU – to reduce the amount of nitrogen in the turbine exhaust stream that enters the downstream condenser section, a high purity low pressure oxygen plant that is commercially available and produces a high purity oxygen (99.5%, volume) product is used. The power requirements were estimated as 359.4 kW / (lb/sec O_2).

- Oxygen / Fuel Compressors – Two multistage intercooled oxygen compressors were used, a six stage unit supplies oxygen at 2500 psia to the gas generator and a three stage unit supplies oxygen at 420 psia to the reheat combustor. A two stage compressor is used for the fuel stream supplied to the gas generator.

- Gas Generator – this section was represented using an ASPEN reactor model. The input streams consisted of natural gas (represented as methane), high pressure steam and high pressure water. The cost estimate was made using information furnished in CES reports with a process contingency of 25% used.

- High Pressure Turbine / Steam generator – Power was generated using a HPT with the exhaust used to generate steam before being sent to the reheat combustor.

- Reheater – oxygen combustor that reheats the process stream using additional methane fuel to raise the temperature to 2600 °F before entering a final series of turbine expanders. Again the cost estimate was based on CES information.

- Intermediate/Low Pressure Turbines – The gas stream has a composition of about 90% steam, 10 % CO_2 with small amounts of nitrogen/argon impurities. Thermodynamic properties used were based on an equation of state for highly non-ideal system (Schwartzentruber-Renon) to accurately represent this stream. Costs for all turbines (HPT,IPT,LPT) were based on using the ICARUS costing software. A 25% process contingency was used.

- Heat Recovery / Condenser – the process stream at 2.1 psia enters a heat exchanger used to generate steam before entering the condenser. Depending on the temperature of available cooling water, different amounts of water can be condensed out. Based on cooling the process stream to 100 °F, approximately 88% of the water is condensed out for recycling.

- CO_2 Compression Process - An intercooled seven stage compression process was used to eliminate any remaining water and to produce a CO_2 product stream at 2100 psia which was cooled to 100 °F and then pumped to 3000 psia for storage. An ICARUS estimate for this section results in a cost of 31500 K$ or approximately $1000/kW. (based on the compressor power). The first stage compressor because of the low inlet pressure (1.9 psia) is beyond most available single train equipment and will require several trains of equipment.

The Aspen Plus® simulation and the cost estimate yielded the results listed in Table 9 above. The overall process efficiency of 48.3%, the total capital requirement of $975/kW, and the COE estimate of 49.2 $/MW-Hr indicate poorer performance when compared with the reference NGCC plant that included CO_2 capture. (i.e. 49.9%, 911 $/kW , 46.4 $/MW-Hr) . The oxygen plant, oxygen compressors and CO_2 section account for over 55% of the equipment costs. CES has efficiency estimates that appear to be approximately 2 percentage points higher for these conditions and higher estimates based on using conditions that appear to be either questionable such as 3200 °F turbine inlet temperatures or low exhaust pressures of 0.65 psia which will increase the cost of the CO_2 compression process. Additional simulation results are provided in Appendix A and the COE cost spreadsheet is provided in Appendix B.

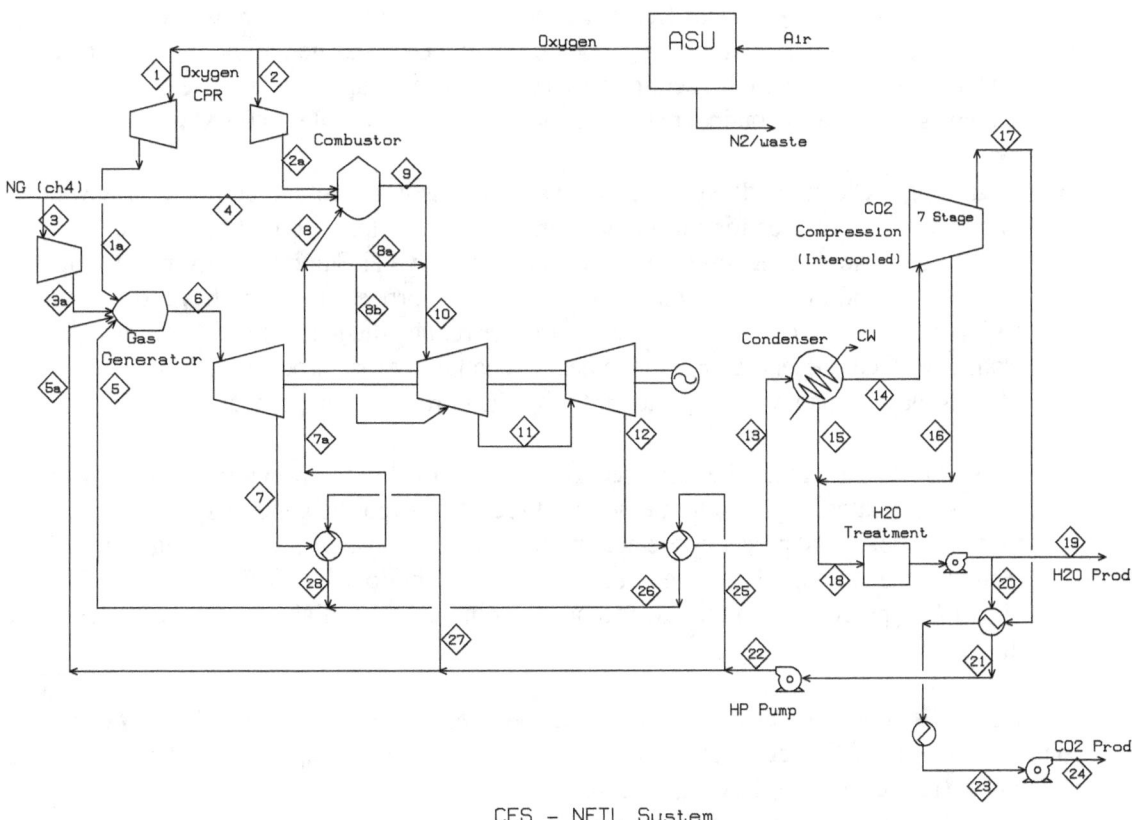

Figure 19. CES – Rocket Engine Steam Cycle – 400 MWe – Natural Gas

II-2.2 Clean Energy Systems (CES) - Coal Syngas System

An Aspen Plus® simulation, Figure 20, was developed to evaluate the performance and cost of the proposed CES process when fueled with a coal syngas. The representation for the natural gas system was combined with sections of a Destec IGCC process based on HGCU. The major sections included were:

- Cryogenic ASU – the same high oxygen purity system was used and the capacity adjusted to provide oxygen for the gasification area.

- Destec Gasification / Syngas Cooler – the gasifier was operated at approximately 1000 psia. The higher pressure gasifier was used to provide the highest pressure deemed feasible for the fuel stream being generated for the CES gas generator. The syngas cooler was integrated into the CES section to serve as a steam superheater. The coal flowrate used was adjusted to obtain a net power output of approximately 400 MWe.

- Coal Syngas Cleanup – the gasifier/syngas cooler as in the reference IGCC case was followed with cyclones for particulate removal and a chloride guard bed . The transport desulfurizer / acid plant approach were used to remove H_2S and COS from the syngas stream. Depending on the requirements of the CES process this may have to be augmented with additional guard bed to further reduce the sulfur level. The gas stream from the HGCU regenerator enters a heat exchanger before proceeding to the acid plant. This exchanger also was integrated into the CES process to superheat steam.

- CES process - Includes the same sections as described in the previous sections with the feed water heaters adjusted to include heat recovery from the gasifier syngas cooler and from the cooler that precedes the acid plant. Due to the use of the coal syngas instead of methane, the amount of CO_2 generated approximately doubles. This is reflected in a higher CO_2 percentage in the low pressure turbine exhaust of about 18% versus 10% for the natural gas case.

The ASPEN Plus simulation and the cost estimate yielded the results listed in Table 9 above. A comparison with the Destec reference case that included CO_2 capture indicated slightly better performance: (reference case shown in brackets)

Overall Process Efficiency :   41.4% [40.1%] ,
Total Capital Cost $/kW   :   1768   [1897] ,
COE $/MW-Hr   :   49.2   [46.4].

CES has efficiency estimates that were based on using a Texaco gasification process that appear approximately the same as these results for the process efficiency. Details of these two simulations have been provided to CES (Larry Hoffman) and are provided in Appendix A and Appendix B.

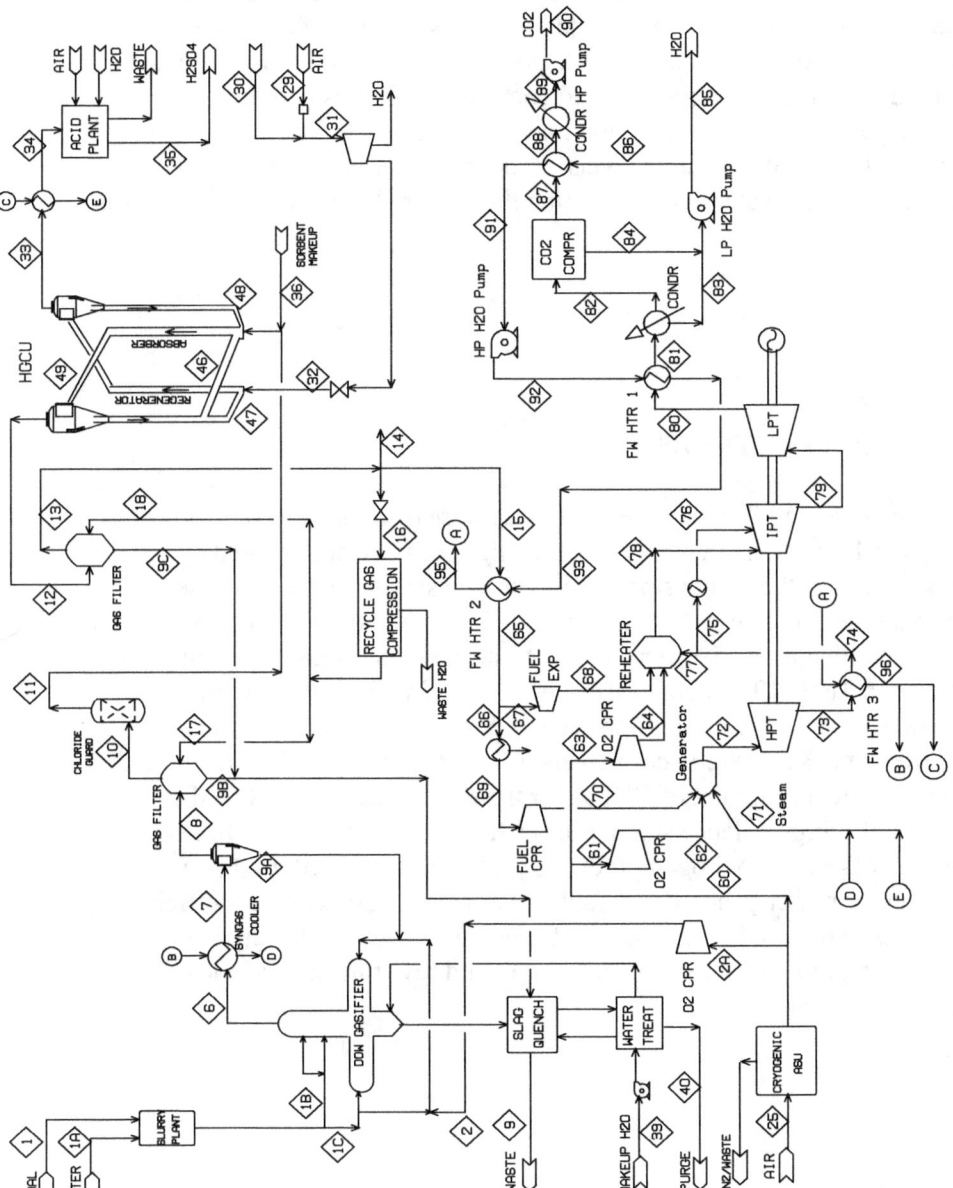

Figure 20. CES – Rocket Engine Steam Cycle – 406 MWe – Coal Syngas.

47

II-2.3 Clean Energy Systems (CES) - Summary

The CES Rocket Engine Steam cycles based on either natural gas or coal syngas do not appear to be able to reach the performance levels of the Vision 21 program. Additionally, considerable effort both in research and funding is anticipated to develop the gas generator and the ultra high pressure/temperature turbines. The oxygen combustion process envisioned increases the oxygen required significantly when compared with an oxygen blown IGCC process. This leads to some projected improvement in performance and cost if the cryogenic ASU is replaced with a membrane process (ITM or OTM) for oxygen production. Another problem area is the large compression cost for the first stage of the carbon dioxide recovery system resulting from the low exhaust pressure of $0.65 - 2.1$ proposed by CES. The Aspen Plus® simulations also assumed that both the gas generator and reheater combustor could combust the fuel using near stoichometric amounts of oxygen. Some consideration may be warranted to increasing the low pressure exhaust temperature to near atmospheric levels, recovering energy by generating steam for injection and then condensing the water out and starting the carbon dioxide compression from this higher pressure point.

II-3. HYDROGEN TURBINE CYCLES

As an alternate approach for achieving CO_2 capture, two cases were developed using a power cycle based on the gas turbine being fueled with hydrogen. High pressure air supplied by the compressor section was still used in the combustor. The hydrogen stream in the first case is based on using steam reforming of natural gas (methane used for simulations) and for the second case on using an IGCC process that uses coal. The results from the Aspen Plus® simulations and the COE analysis are shown in Table 10. In both cases, the gas turbine fueled by hydrogen produces 269 MWe of power. The CO_2 compression section power requirements are (as expected) significantly different (13.5 MWe in case 1 versus 31.6 MWe in case 2) due to the coal case generating more than double the amount of CO_2 as for the natural gas case. Flow diagrams are provided with material and energy balance summaries in Appendix A and the COE results are in Appendix B. For both cases, the hydrogen produced probably is bettered used as a chemical product rather than for power generation. Table 10 indicates both a process efficiency based on the amount of methane required in the steam reformer and based on the amount of hydrogen used. An alternate process that uses less methane would result in an efficiency between these two values.

Table 10. Hydrogen Turbine Power Cycles.

POWER SYSTEM	HYDROGEN TURBINE (HT)	
Generation Cycle	HT (H2 FROM SMR) (CO2 CAPTURE)	HT / COAL Destec HP (E-Gas) HGCU (CO2 CAPTURE)
Net Power MWe	413.1	375.3
Net Plant Efficiency	64.4 (H2)	38
% LHV	42.9 (NG)	
Total Capital Requirement	1323	1909
$ / KW		
Cost of Electricity	63.5	53.6
Constant $ / MW-hr		
NOx emissions	0.161	0.177
lb/MW-hr		
Sox emissions	---	0.046
lb/MW-hr		
CO2 Production		
lb/MW-hr		
a) Emitted to atmosphere	---	---
b) Sequesterable	719	1731
Footprint (battery limits)	472	1445
sq ft/MW		

II-3.1 Hydrogen Turbine Cycles – Natural Gas Case

This case was developed by modifying the NGCC reference simulation (see I-2.1) to use hydrogen in place of natural gas as the fuel for the gas turbine. The required hydrogen was assumed to be supplied by a steam methane reformer / hydrogen purification process. (commercially available process [23]). The hydrogen purification uses pressure swing absorption and the CO_2 is recovered by extending the process to include a vacuum swing absorption step. The CO_2 captured was then compressed to a high pressure (2100 psia) to enable

49

sequestration as a liquid product. (The economic analysis does not assume a value for this product or include a transportation charge for disposal.) The steam generated in the SMR was integrated into the combined cycle process to recover additional power. The net power generated was calculated based on the simulation results for the gas turbine, steam turbine, CO_2 captured and a literature estimate for the SMR process. The process efficiency was calculated using the net power generated using both the hydrogen used (in the gas turbine) and the methane used (in the SMR) to generate this hydrogen.

Emissions were calculated for CO_2 based on the natural gas (methane) used in the SMR as fuel. The NOx was estimated based on 9 ppmv for the gas turbine section added to an estimate for the SMR plant.

The COE cost analysis relied on the NGCC reference case augmented by the cost of the SMR plant and the CO_2 compression section. The footprint (battery limits) of the NGCC reference case was similarly increased by an estimate for the SMR and CO_2 recovery equipment. For costing, the overall plant site was considered to cover 100 acres.

Figure 21 illustrates the simulation model representation, Appendix A contains the material and energy balance summaries and Appendix B contains the COE spreadsheet summary.

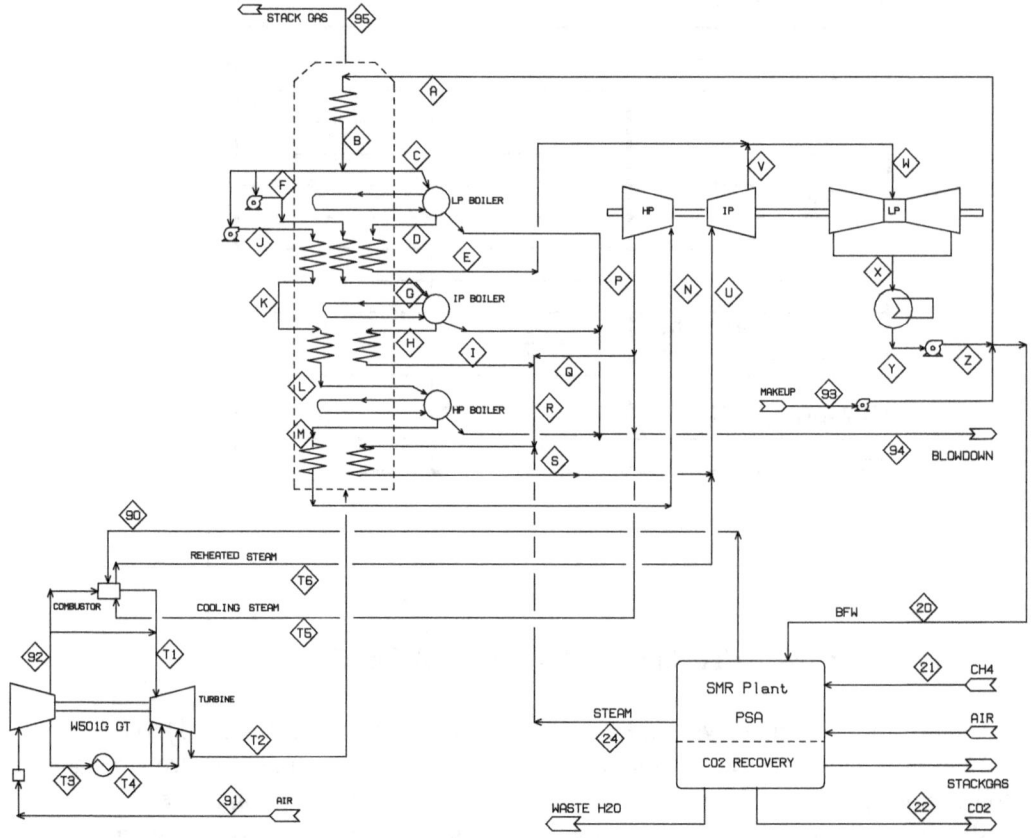

Figure 21. Hydrogen Turbine Cycle – Natural Gas

II-3.2 Hydrogen Turbine Cycles – Coal Case

The Aspen Plus® simulation model was developed by modifying the simulation developed and described above (see section II-1.3) for the Hydraulic Air Compression (HAC) case with CO_2 capture. The major plant sections: high pressure Destec Gasifier, ASU, HGCU, Acid plant, HSD, H_2 stream HRSG, CO_2 stream HRSG , CO_2 compression, steam turbines are retained from the HAC case. The necessary changes are:

- the insertion of a section for the hydrogen powered "G" gas turbine , HRSG and steam cycle to replace the hydraulic compression/recuperator sections in the HAC case. This is the section equivalent to the above natural gas case (see II-3.1, Figure 21).

The resulting process is shown in Figure 22 (similar to Figures 16, 17). This case indicates a decrease in process efficiency to 38 % (LHV) compared to the 42.9% determined for the natural gas fueled hydrogen turbine cycle. Again as a power plant this case appears to have no hope of meeting Vision 21 goals. Alternately, the gas turbine and steam cycle sections can be omitted and the process viewed as producing hydrogen and power. The heating value of the hydrogen (100%) is then used to calculate a combined heat and power efficiency for which the Vision 21 goal is 85 – 90% (HHV) based on coal fuel [24]. The present case based on a hydrogen production (45384 lbs/hr) and the remaining net power would yield the following:

CHP eff. (HHV) =   100% * (H_2 heating value + net power*3414) / (coal heating value)

=   79.4%

This is again below the Vision 21 objectives.

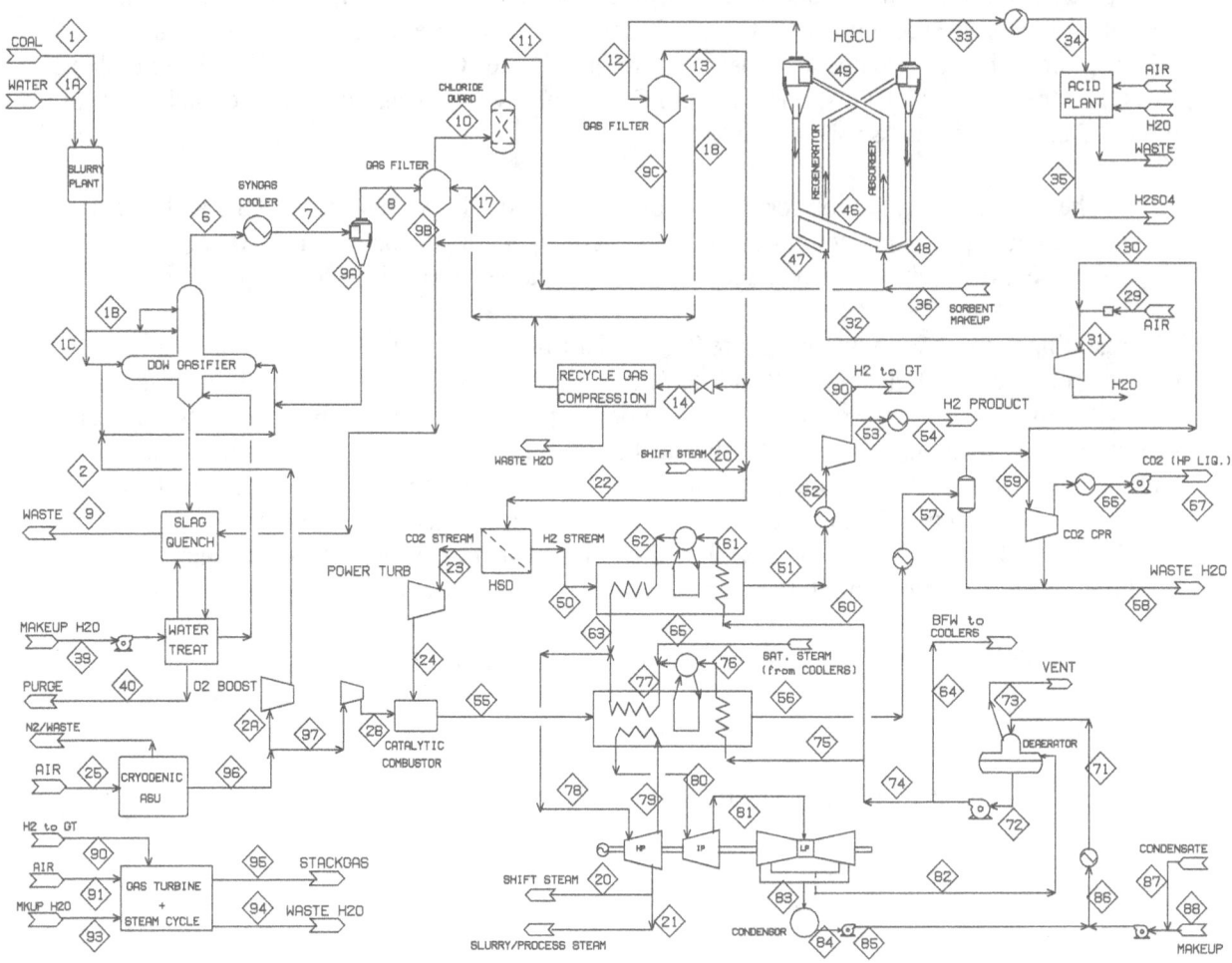

Figure 22.   Hydrogen Turbine Cycle – Coal

II-3.3 Hydrogen Turbine Cycles – Summary

The hydrogen turbine cycles as summarized in Table 10 have poorer performance and higher cost when compared with reference cases. The natural gas fueled case has an efficiency of 42.9% (LHV) and a COE of 61.2 $/MW-Hr. The NGCC reference case that uses an amine process for CO_2 capture has a higher efficiency of 49.9% and a COE of 46.4. The coal fueled case efficiency of 38.0% is approximately the same as the Destec/CGCU reference case results of 40.1% .

An alternate hydrogen turbine cycle has been proposed for coal gasification systems that rely on hydrogen combustion with oxygen [25]. Steam is injected in the combustors in a manner that is somewhat similar to the CES systems described in II-2. The coal syngas generated by gasification is shifted and sulfur compounds and CO_2 removed using the RECTISOL absorption process and sulfur recovered in a CLAUS/SCOT section. The hydrogen produced is split between a high pressure combustor and a reheat combustor between two turbine expander sections. A HRSG is used to generate steam before the flue gas (essentially steam) is expanded in a low pressure turbine section. The process projects efficiencies of approximately 50% (HHV) which includes CO_2 compression to 80 bar (1160 psia).

II-4. HYBRID - TURBINE / FUEL CELL CYCLES

A Hybrid power system configured as a combined cycle based on using a high temperature Fuel Cell and a Gas Turbine holds promise for approaching the efficiency goals of the Vision 21 program. DOE is currently sponsoring a number of programs both to develop fuel cells, to compare different hybrid concepts and to evaluate related technical issues [26]. Major hurdles also included reducing the cost and size of the fuel cell modules to make hybrid systems available for generating electrical power in commercial power plant sizes > 100 MWe. The current report considers the systems summarized in Table 11 and are based on using Solid Oxide Fuel Cells (SOFC). The efficiencies shown are based on using currently available components and projected performance for the SOFC modules. Modest improvements in turbine and/or fuel cell performance would probably result in these systems obtaining the Vision 21 goals of 75% (LHV) for natural gas systems and 60% (HHV) for coal systems.

Table 11. Hybrid Turbine/Fuel Cell

POWER SYSTEM	HYBRID CYCLE (HYB)			
Generation Cycle	Natural Gas Hybrid Turbine- SOFC Cycle	HYB / COAL Destec (E-Gas) HGCU "G" GT / SOFC (NO CO2 CAPTURE)	HYB / COAL Destec HP (E-Gas) HGCU / HSD "G" GT / SOFC (CO2 CAPTURE)	HYB / COAL Destec (E-Gas) OTM / CGCU "G" GT / SOFC (NO CO2 CAPTURE)
Net Power MWe	19	643.6	754.6	675.2
Net Plant Efficiency % LHV	67.3	56.4	49.7	57
Total Capital Requirement $ / KW	1476	1508	1822	1340
Cost of Electricity Constant $ / MW-hr	53.4	41.1	48.8	38
NOx emissions lb/MW-hr	0.0132	0.107	0.093	0.101
Sox emissions lb/MW-hr	---	0.005	0.004	0.014
CO2 Production lb/MW-hr				
a) Emitted to atmosphere	661	1254	101	1237
b) Sequesterable			1323	
Footprint (battery limits) sq ft/MW	1120	1310	1408	1388

II-4.1 Hybrid - Turbine/Fuel Cell Cycles – Natural Gas Case

The results for this case were obtained from a report "Pressurized Solid Oxide Fuel Cycle/Gas Turbine Power System" by Siemens Westinghouse / Rolls-Royce Allison for the DOE. (DE-AC26-98FT40355 , February 2000) [27]. (The reported performance was verified using an Aspen Plus® simulation).

The DOE report describes the development of a conceptual design for a pressurized SOFC/GT power system that was intended to generate 20 MWe with at least 70% efficiency. The system shown, Figure 23, designated the HEFPP system cycle (High Efficiency Fossil Power Plant) integrates an intercooled, recuperated, reheated gas turbine with two SOFC generator stages. One SOFC stage operates at high pressure, and generates power as well as providing all heat needed by the high pressure turbine. The second SOFC generator operates at a lower pressure, generates power, and provides all heat for the low pressure reheat turbine. The system is projected to have an efficiency of 67.3% (LHV).

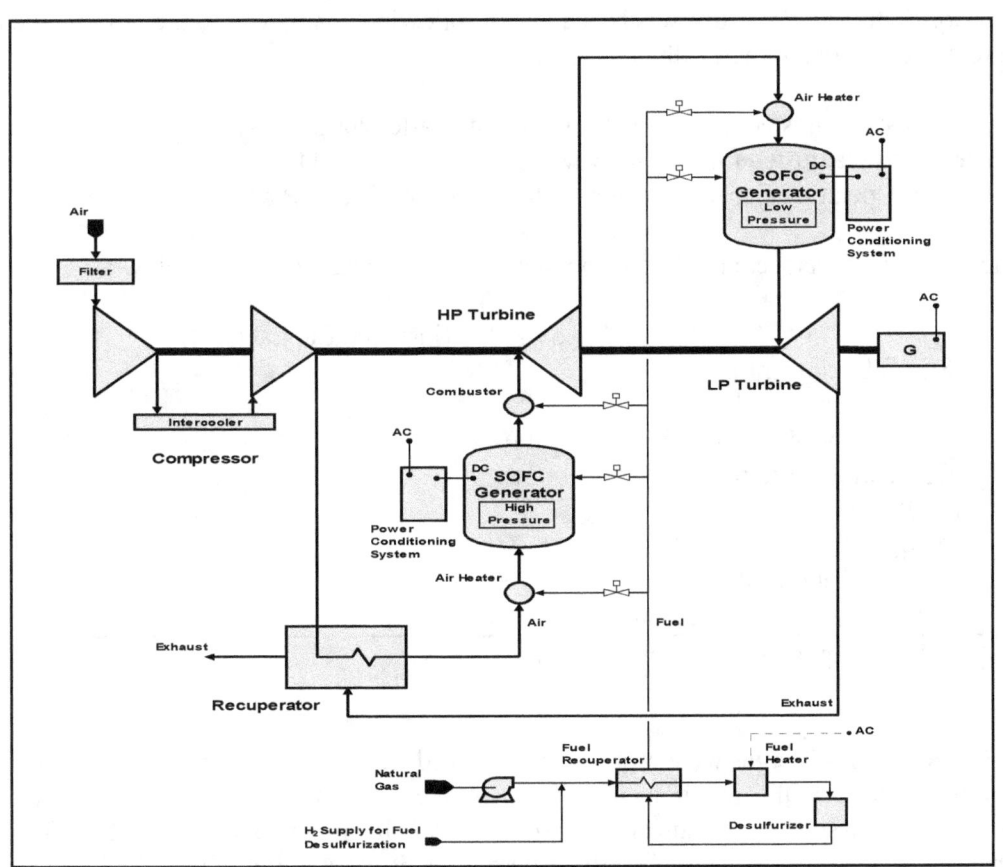

Figure 23. High Efficiency Fossil Power Plant Cycle (HEFPP)

The following design conditions are summarized from the report:

Approximate Power Generation : 15 MWe from SOFC and 4 MWe from Gas Turbine.
Fuel – Methane (96%), Nitrogen (2%), Carbon dioxide (2%), Sulfur (4 ppmv)
Air – inlet flow rate to air compressor = 40 lbs /sec, 59 °F, 14.7 psia.
Air Compressor – two stages intercooled , overall pressure ratio = 7:1 ,
        isentropic efficiency = 86.4%.
Recuperator – preheats high pressure air using LP turbine exhaust to about 1126 °F.
HP SOFC – operates at exit conditions of about 1600 °F and 92 psia ,
        required fuel inlet sulfur level 0.1 ppmv , 90% fuel utilization.
HP Turbine – isentropic efficiency = 90.7 % , inlet temperature = 1600 °F.
LP SOFC - inlet conditions of 46 psia and 1300 °F , exhaust at 40 psia and 1600 °F,
        90% fuel utilization.
LP Turbine - exhaust at 1197 °F and 15.5 psia, isentropic efficiency = 91.3%.

Emissions - CO_2 = 661 lbs / MW-hr , NOx 0.013228 lbs / MW-hr

The Cost Estimate developed was based on the following changes from the report which were made for consistency with other COE estimates:

Plant Costs - adjusted from 1998 basis to first quarter 2002 basis
Fuel Costs – adjusted from $3.0 / MMBU to $3.2 / MMBTU.
Annual Operating Capacity Factor – adjusted from 92% to 85%

The following costs were modified from the report to the values shown below :

	$ / KWe (installed Capital Cost)
SOFC Equipment	
- Generator	486
- Power Conditioning	110
Gas Turbine Equipment	218
BOP	267
Site Prep, M & E	72
Overhead and Profit	300
Spare parts, startup, & land	23
Total Capital Requirement	$1476 / KWe

Several battery limits designs were proposed that ranged from 0.5 – 0.6 acres. The battery limits are dominated by the SOFC requirements. Siemens Westinghouse / Rolls-Royce Allison project that an optimized system can obtain an efficiency > 70%. It should be noted that the SOFC performance has been estimated with perhaps an optimistic assumption of 90% fuel utilization.

A coal fueled version of this system for a nominal 500 MWe size plant has been previously formulated and projected to have an efficiency of 59% (HHV) [26]. Additional studies are currently being sponsored by NETL for systems based on Shell and Texaco gasifiers. [28]. The results will be available in 2003.

II-4.2 Hybrid - Turbine/Fuel Cell Cycles – Coal Cases

Aspen Plus® simulations were developed for two new cases based on using coal and these results are combined with a third case from an earlier study [29] and summarized in Table 12. All cases were based on using a Destec gasifier, a W501G gas turbine, and a SOFC. The syngas generated was split with 58% sent to the SOFC and the remaining 42% sent to the gas turbine combustor. The coal flowrate was adjusted so that the power produced by the gas turbine was approximately 275 MW for all three cases. Shifting more syngas to the fuel cell will increase efficiency but additionally increase the COE because of the increase in the number of fuel cell modules required. (A capital cost of $800/KW was assumed for the fuel cell section).

Table 12. SUMMARY - SIMULATION FOR COAL SYNGAS HYBRID POWER SYSTEMS

CASE	CO$_2$ Capture	GAS CLEANUP	TURBINE FUEL	% SYNGAS TO SOFC	(POWER IN MWe)					EFF % LHV
					NET POWER	GAS TURBINE	STEAM TURBINE	SOFC	MISC/AUX	
1	NO	HGCU/ZNO	SYNGAS	58%	643.6	276.1	207.7	221.4	61.5	56.4
2	YES	HGCU/ZNO	H$_2$	58%	754.6	272.5	226.1	324.1	68.2	49.7
3	NO	CGCU	SYNGAS	58%	675.2	272.7	189.8	254.4	41.8	57

II-4.2.1 Hybrid - Turbine/Fuel Cell Cycles – Coal Cases – Case 1 (No CO$_2$ Capture)

Case 1 was developed based on making the following modifications to the reference Destec / HGCU case (see Table 2, Figure 8) and does not include CO$_2$ capture:

- An additional zinc oxide guard bed is added to the HGCU section to reduce the sulfur content of the cleaned fuel gas to acceptable levels for use in the SOFC. (assumed 1-5 ppmv was acceptable and obtainable).

- The SOFC section was added using a previously developed fuel cell model [30]. A fuel utilization of 85% was assumed. The anode and cathode exit streams are combined and the remaining fuel combusted to raise the temperature > 2000 °F. This stream is used to preheat the cathode inlet stream and then routed to the gas turbine combustor.

57

- The gas turbine compressor outlet provides 50% of the air required by the ASU and all the air required by the HGCU regenerator as in the reference case. The remaining air is combined with a nitrogen recycle from the ASU and sent to the cathode preheater before entering the fuel cell.

- The cleaned fuel gas is split between 58% entering the fuel cell and 42% sent to the gas turbine combustor.

- The steam cycle design is the same as for the reference case.

In Figure 24, the resulting hybrid GT/SOFC is shown. Appendix A has material and energy balances and Appendix B contains the COE spreadsheet results.

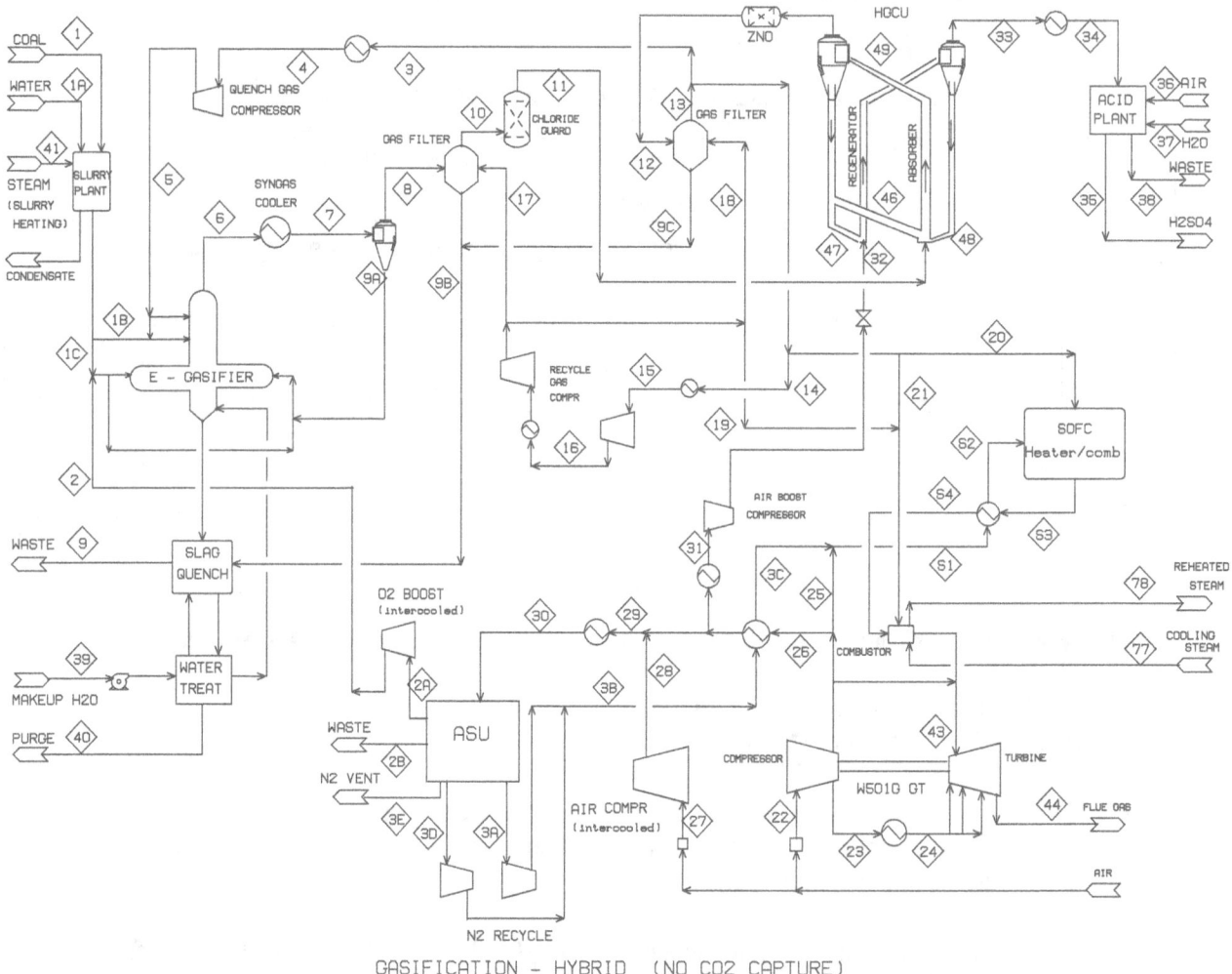

GASIFICATION - HYBRID (NO CO2 CAPTURE)

Figure 24. Case 1. Hybrid GT/SOFC – Coal Syngas – No CO$_2$ Capture

II-4.2.2 Hybrid - Turbine/Fuel Cell Cycles – Coal Cases – Case 2 (CO_2 Capture)

Case 2 was developed based on making the following modifications to the Hydrogen Turbine Coal Cycle case (see Table 10, Figure 22) and includes CO_2 capture:

- An additional zinc oxide guard bed is added to the HGCU section to reduce the sulfur content of the cleaned fuel gas to acceptable levels for use in the SOFC. (assumed 1-5 ppmv was acceptable and obtainable).

- The SOFC section was added using a previously developed fuel cell model [30]. A fuel utilization of 85% was assumed. The cathode exhaust is used to preheat the cathode inlet stream (high pressure air from the gas turbine) and returns to the gas turbine combustor. The anode stream containing unspent fuel is expanded in a power turbine and combined with the CO_2 rich stream from the HSD (hydrogen separation device) and the combined stream enters a catalytic combustor.

- The gas turbine compressor outlet provides 50% of the air required by the ASU and all the air required by the HGCU regenerator. The remaining air is sent to the cathode preheater before entering the fuel cell.

- The cleaned fuel gas is split between 58% entering the fuel cell and 42% sent to the HSD. The fuel sent to the HSD is used to produce hydrogen for the gas turbine.

- Nitrogen is recycled from the ASU to the gas turbine combustor after being preheated in two heat exchangers. The first exchanger uses the hydrogen exhaust stream from the HSD and the second exchanger uses the exhaust from the catalytic combustor.

- Heat Recovery Steam Generators (HRSG) are used to recover available heat in the turbine exhaust, gasifier syngas cooler and the catalytic combustor exhaust. The generated steam is used for power generation and supplying steam for the HSD (shift reaction) and for heating in the slurry plant.

- The CO_2 capture uses the same approach as in the Hydrogen Turbine Case with a high pressure liquid stream produced.

In Figure 25, the resulting hybrid GT/SOFC is shown. In Figure 26, the steam cycle is shown. Appendix A has material and energy balances and Appendix B contains the COE spreadsheet results. As in Case1, the efficiency will improve as more fuel is sent directly to the fuel cell. The fuel split assumed (58% to the fuel cell) was made due to the high capital cost ($800/KW) used for the SOFC modules and the desire to use the modified W501G gas turbine.

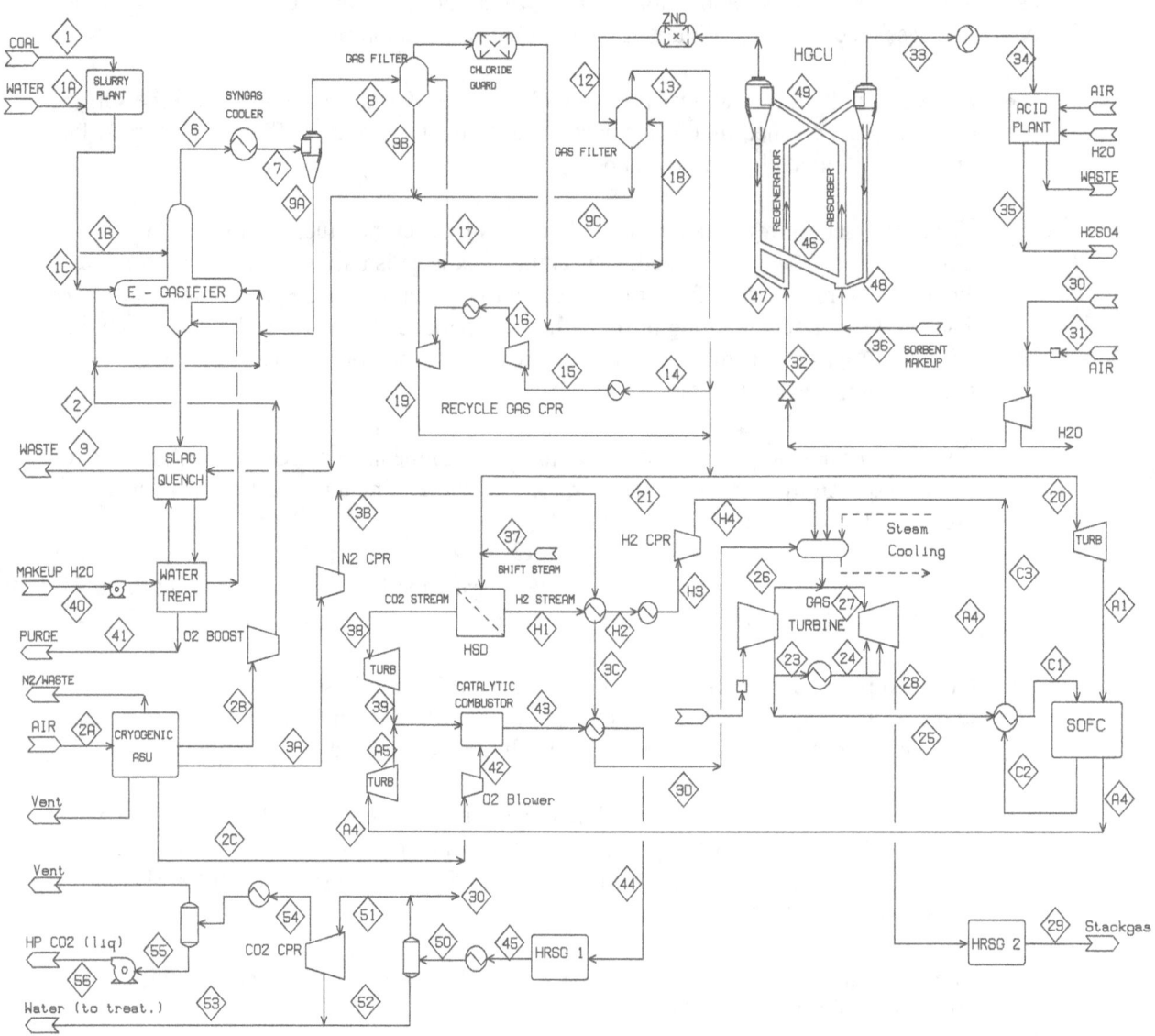

Figure 25. Case 2 . Hybrid GT/SOFC – Coal Syngas – CO₂ capture

60

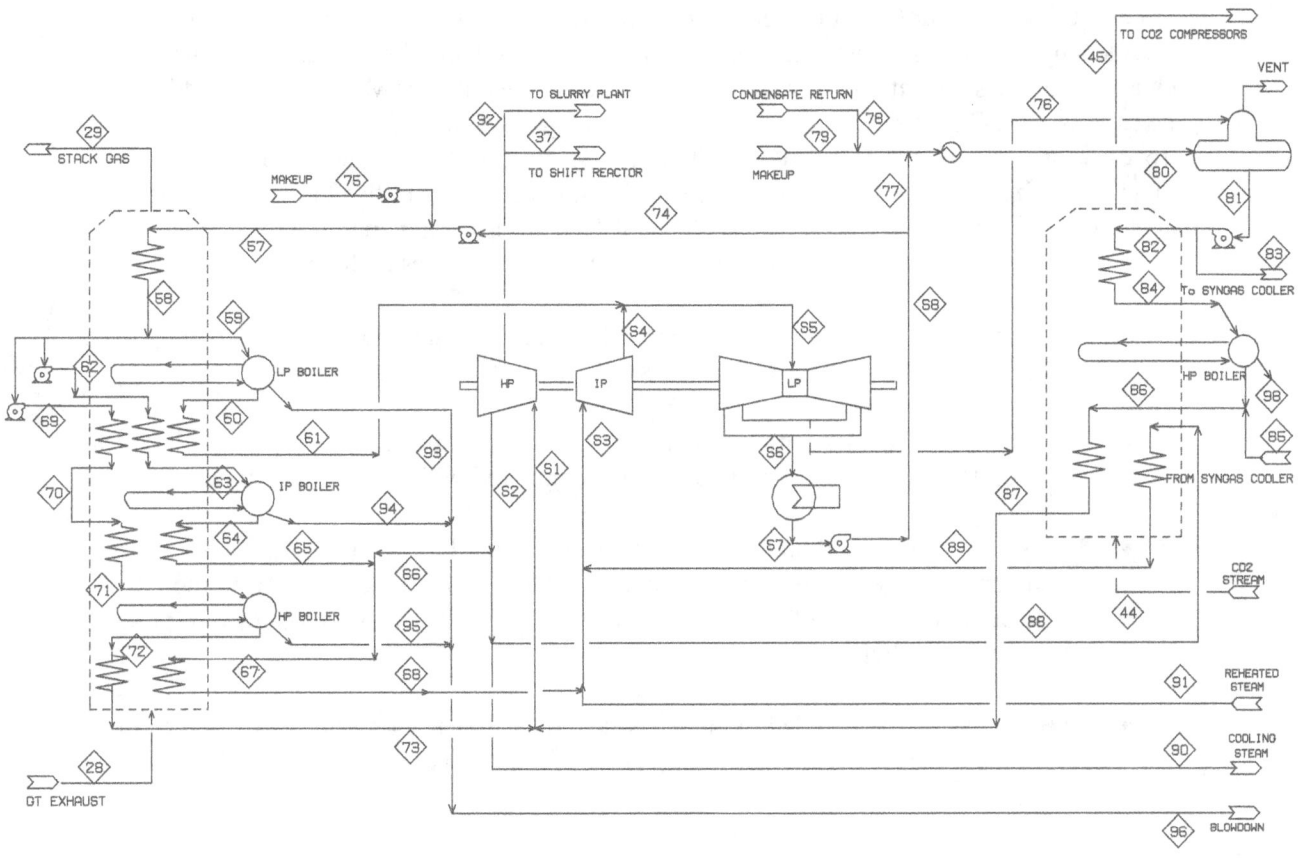

Figure 26. Case 2 . Hybrid GT/SOFC – Coal Syngas – CO$_2$ capture

II-4.2.3 Hybrid - Turbine/Fuel Cell Cycles – Coal Cases – Case 3 (No CO_2 Capture)

Case 3 was initially developed as part of a CRADA between NETL and Praxair [29]. This CRADA examined replacing the cryogenic ASU for oxygen production with a membrane process (OTM) in a number of power plant schemes. Case 3 is included in the present report to provide a hybrid that integrates the SOFC with both the ASU and the gas turbine. This takes advantage of the similarity in operating temperature between the OTM and SOFC. Additionally the case uses a CGCU (RECTISOL) to clean the fuel gas to low sulfur levels. In Figure 27, the process is shown. Details of the SOFC/OTM process are confidential and were provided to NETL as a "black box" as shown on the flow diagram. (The Aspen Plus® model developed used a combination of intrinsic reactor models (RGIBBS) and separation operations to obtain approximately the information furnished by Praxair.) Key features include:

- A commercially available process, RECTISOL, is used instead of the HGCU approach used in Case 1 and Case 2 to remove sulfur from the fuel stream.

- The ASU is based on an advanced process under development that is projected to have lower costs and lower energy requirements compared with cryogenic oxygen plants.

- The SOFC is integrated both with the gas turbine and the ASU (OTM). The combined air stream from the gas turbine and supplemental compressor and fuel from the gasification unit are sent to the SOFC/OTM section. The SOFC is assumed to produce power at 50% efficiency.

- CO_2 capture is not included. It is expected that modifications are possible that would result in a CO_2 rich stream. However, an efficiency penalty of 6-7% would be projected as in other cases.

- Developers of ASU processes expect performance and costs to improve over the assumptions used for the present case that yields a 57% (LHV) efficiency.

- The steam cycle integrates available heat into a three pressure level steam cycle similar to Case 1. Steam is provided for the CGCU and Slurry plant sections.

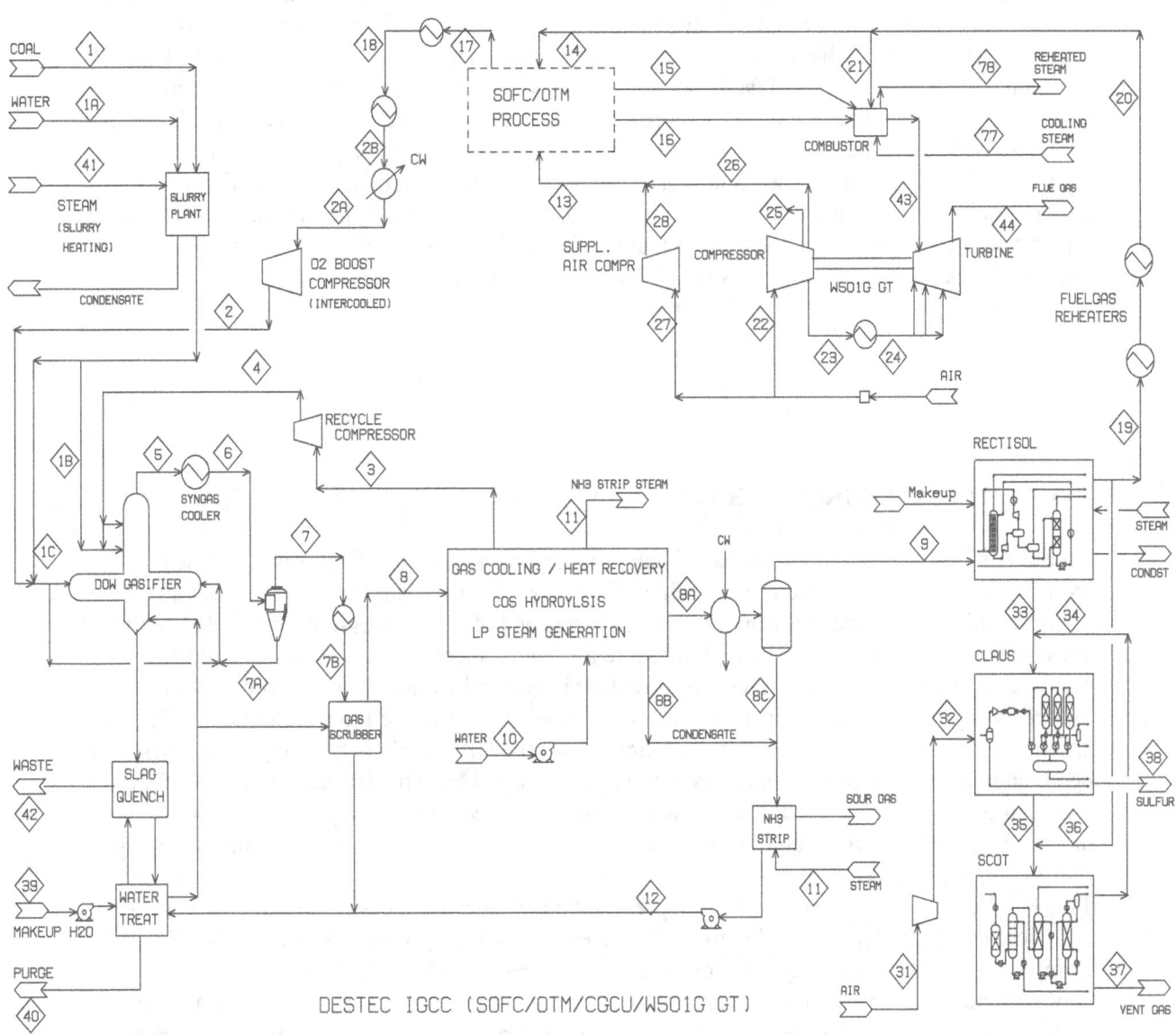

Figure 27. Case 3 . Hybrid GT/SOFC/OTM – Coal Syngas – No CO$_2$ Capture

II-4.3 Hybrid - Turbine/Fuel Cell Cycles – Summary

The Hybrid cases based on using a combination of a SOFC and turbines resulted in the highest efficiencies obtained for the systems included in this report. The systems considered as summarized in Table 11 have efficiencies that are approaching the goals of the Vision 21 program of 75% (LHV, natural gas) and 60% (HHV, coal). The coal cases considered in this report will have higher efficiencies as more of the fuel is sent directly to the fuel cell. The fuel split assumed was primarily made because of the high cost currently projected for fuel cells and the use of the "G" turbine. The natural gas case uses turbines with relatively low firing temperatures and performance will increase with different choices for the turbines. However, this optimistic feeling is made assuming that the fuel cells performance can be demonstrated for large modules and that the cost ($/KW) is drastically reduced.

II-5. HUMID AIR TURBINE (HAT) CYCLES

Humid Air Turbine (HAT) cycles have been proposed for a number of years as a means for reducing costs when compared to Combined Cycles (CC). A typical HAT cycle uses a high pressure ratio gas turbine (pressure ratio > 50) composed of a high pressure intercooled shaft and a low pressure power shaft . The high pressure air from the compressor is cooled and then humidified in an air saturator. The humidified air is heated in a heat recovery section that uses the turbine exhaust before entering the turbine combustor. Compared to a combined cycle , the argument is usually made that while the efficiency of the HAT cycle is typically lower by several percentage points that the advantage is in the cost being lower. This is based on the HAT cycle claiming that eliminating the HRSG/Steam Cycle reduces cost more than the added cost of a more expensive gas turbine and the addition of the air saturator and a number of heat exchangers.

Two HAT cycles were considered in the present study (natural gas case, and a syngas case) based on a turbine design provided to NETL by Pratt & Whitney Power Systems (PWPS) [31]. Comparisons to the reference cases for NGCC and a Destec/CGCU IGCC indicated approximately the same efficiencies and higher costs for the HAT cycles. PWPS would not provide a cost estimate for the high pressure ratio gas turbine since it's currently in the research and development stage. This cost was estimated based on information from the EG&G Cost Estimating Notebook (version 1.11) and are included with the COE spreadsheets for these cases in Appendix B. (The COE results can be easily revised if information becomes available.) Since the two HAT cases developed demonstrated no advantage over reference cases, HAT cycles that include carbon dioxide capture were not considered. NETL is currently funding systems studies (no COE analysis) based on HAT cycles combined with SOFC [28], that demonstrate high efficiencies. However, the efficiency gain found in these studies is due to the use of fuel cells and partly due to optimistic efficiencies assumed for compressors and turbine expanders. HAT cycles in non-hybrid systems appear to have no hope of meeting Vision 21 goals. The results obtained for the two cases in the present study are listed in Table 13.

Table 13. HAT Cycle Summary

POWER SYSTEM	HUMID AIR TURBINE (HAT)	
Generation Cycle	**HAT (PW GT) Natural Gas**	**HAT COAL (PW GT) Destec (E-Gas) CGCU**
Net Pow er MWe	318.7	407.4
Net Plant Efficiency % LHV	57.6	44.9
Total Capital Requirement $ / KW	873	1552
Cost of Electricity Constant $ / MW-hr	47	45.1
NOx emissions lb/MW-hr	0.074	0.071
Sox emissions lb/MW-hr	---	0.353
CO2 Production lb/MW-hr		
a) Emitted to atmosphere	758	1576
b) Sequesterable		
Footprint (battery limits) sq ft/MW	175	811

II-5.1 Humid Air Turbine (HAT) Cycles – Natural Gas

Based on the information provided by PWPS, a natural gas HAT cycle was developed and is shown in Figure 28. The aeroderivative turbine consists of a dual shaft arrangement having an overall pressure ratio of 54.2. Other conditions include an inlet air flowrate of 643.3 lbs/sec and a turbine inlet temperature of 2750 °F. The HAT approach results in the elimination of the HRSG/Steam Cycle of the NGCC and adds several heat exchangers (water heating), an air saturator and a heat recovery section. The heat integration allows the high pressure air stream exiting the saturator to have a moisture content of 19.2%. This plant produces a net power of

318.7 MWe and has an efficiency of 57.6% (LHV). Appendix A contains the material and energy balances and the COE is included in Appendix B.

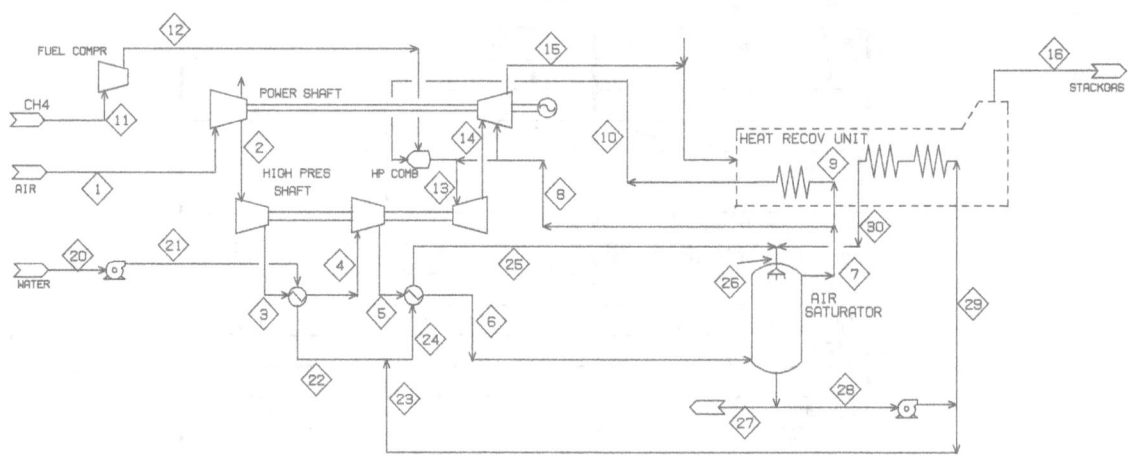

Figure 28. HAT Cycle – Natural Gas - PW Aeroderivative Turbine

II-5.2 Humid Air Turbine (HAT) Cycles – Coal Syngas

An Aspen Plus® simulation model was developed for an Integrated Gasification Humid Air Turbine (IGHAT) based on the following key sections:

- Destec (E-Gas) Gasifier - operates at exit conditions of 1900 °F and 412 psia. Condition and model incorporated from reference Destec/CGCU case.

- ASU – cryogenic oxygen plant (low pressure).

- High Temperature Syngas Cooling - used to both reheat the clean syngas and to heat high pressure water sent to the air saturator.

- Low Temperature Syngas Cooling – includes COS Hydrolysis and heat recovery. Heat recovery used to generate low pressure steam used for the CGCU section stripper and slurry heating. Condenses most of the water from syngas.

- CGCU – used MDEA/CLAUS/SCOT system for sulfur recovery.

- Syngas Compressor / Reheater – compresses and reheats the clean syngas from the MDEA section for use in the gas turbine combustor.

- PW Aeroderivative Turbine – uses the turbine model representation developed for the natural gas HAT case. (Pressure Ratio = 54.2, TIT = 2750 °F)

- Air Saturator – used to humidify the high pressure air from the gas turbine.

- Heat Recovery Unit (HRU) – uses the turbine exhaust to heat the air from the saturator and to heat a portion of the water used in the saturator.

Figure 29 shows a flow diagram for the process which resulted in a net power generation of 407.3 MWe and an overall efficiency of 44.9% (LHV). This is slightly lower when compared to the 46.7% obtained for the reference Destec/CGCU IGCC process. Material and Energy balance summaries are in Appendix A and the COE spreadsheet results are in Appendix B.

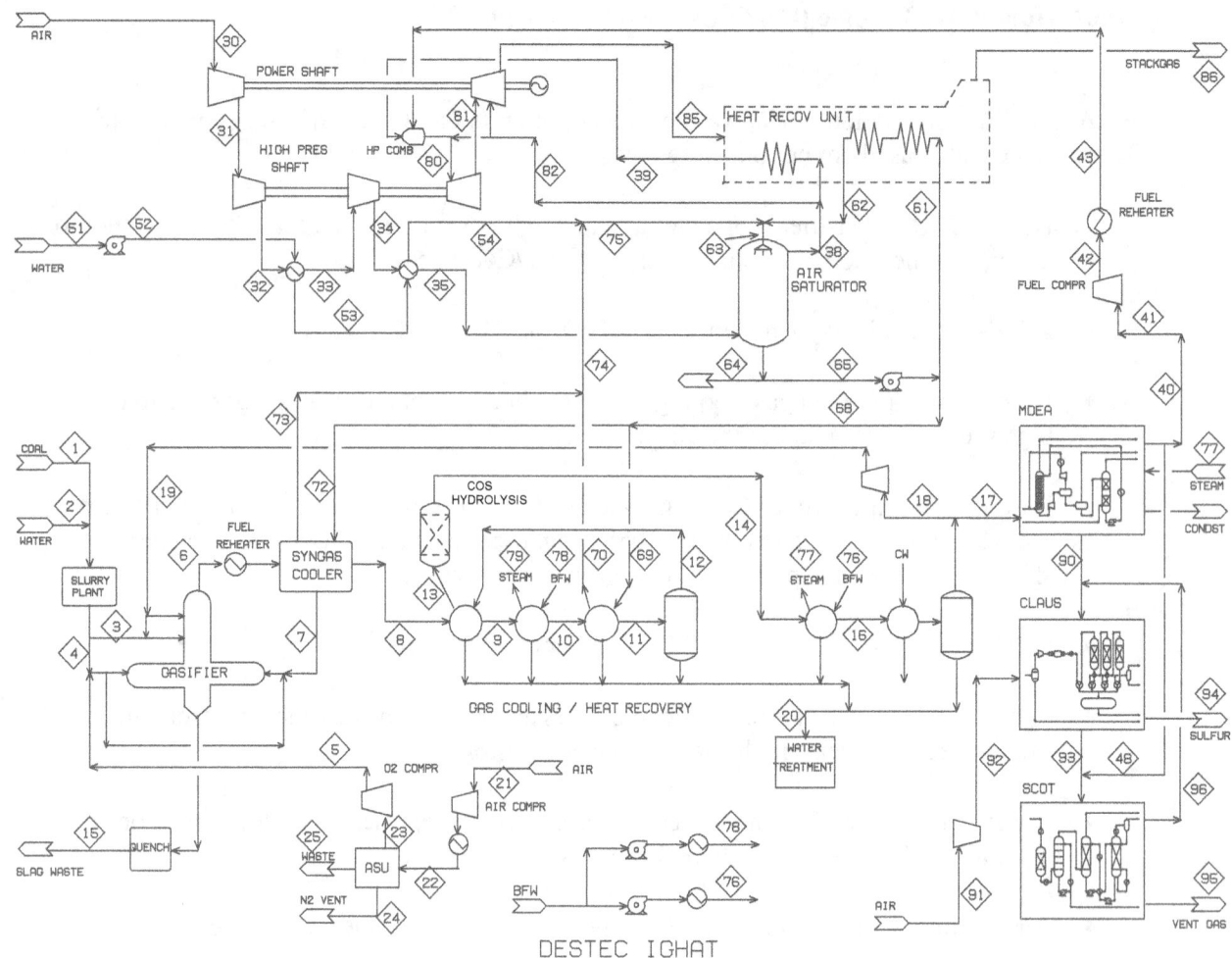

Figure 29. IGHAT – Destec/ CGCU

II-5.3 Humid Air Turbine (HAT) Cycles - Summary

HAT cycles produced efficiencies that were only comparable to corresponding reference combined cycles (NGCC, IGCC). HAT cycles without the addition of a fuel cell and the resulting conversion to a hybrid cycle will not be able to achieve anywhere near Vision 21 objectives. In Table 14 following, a summary is provided of key conditions used and a comparison with simulation results provided by PWPS [31] for a modified HAT cycle based on a TEXACO gasifier. This case uses a small steam cycle and results in a lower moisture content for the humidified air when compared to the Destec HAT cycle. The efficiency is somewhat higher but still significantly below Vision 21 objectives. Systems studies that include hybrid HAT cycles are currently being funded by NETL [28].

Table 14. Comparison with P&W hybrid system and NETL IGHAT Cycle

	P&W IGHAT	NETL IGHAT	NETL NGHAT
Fuel	Syngas (TEXACO Gasifier)	Syngas (DESTEC Gasifier)	Natural Gas (CH4)
Gas Turbine:			
- Pressure Ratio :	54.2	54.2	54.2
- Inlet Air (lbs/sec):	643.3	643.3	643.3
- TIT (^{0}F) :	2750	2750	2750
- weight % Moisture : (air to Combustor)	17	28.1	19.2
Results			
Power (MWe)			
- Gas Turbine	359.9	457.6	326.5
- Steam Turbine	69.6		
- Expander	5		
- Total Gross	434.5	457.6	326.5
- Misc & Aux	51.4	50.2	7.8
- Net Power	383.1	407.4	318.7
Efficiency %			
- HHV	46	43.3	51.9
- LHV	47.7	44.9	57.6

References

1. M. M. Shah , L. Bool, and W. W. Shelton, "Oxygen Transport Membranes (OTM) , CRADA (99-FO32), Task 3: OTM Integrated Combustion Processes, CRADA between Praxair, Inc. and NETL , DOE. April 2001.

2. T.L. Buchanan, M.R. DeLallo, H.N. Goldstein, G.W. Grubbs and J.S. White, "Market-Based Advanced Coal Power System," (Section 3.1, Pulverized Coal-Fired Subcritical Plant , 400 MWe), Final Report, December 1998, Prepared For: The United States Department of Energy, Office of Fossil Energy, Contract No. DE-AC01-94FE62747, BY: Parsons Infrastructure & Technology, Reading, Pennsylvania.

3. Praxair, Inc. - communication from M. Shah , (commercially available costs), to NETL , 1/15/2001.

4. G. T. Rochelle, University of Texas at Austin, "Carbon Dioxide Capture by Absorption with Potassium Carbonate" , NETL funded project , March 6, 2002.

5. Private Communication - R. Birnbarw (DOW Chemical) to NETL, April 11, 2002.

6. "Gas Turbine World 2001"

7. "Gasification Technologies – Gasification Markets and Technologies – Present and Future – An Industry Perspective" (DOE report, July 2002)

8. NETL/Gasification Technologies team website (System Studies FY2000, EG&G) : http://www.netl.doe.gov/coalpower/gasification/system/destx3x_.pdf

9. NETL/Gasification Technologies team website (Wabash River Coal Gasification Repowering Project): http://www.netl.doe.gov/coalpower/gasification/projects/system

10. NETL/Gasification Technologies team website (Global Energy, Nexant) : http://www.netl.doe.gov/coalpower/gasification/projects/systems/docs/40342R01.PDF

11. Private Communication – "SSK , "Sulfur Tolerant Shift Catalyst", Jens Houken (Haldor Topsoe, Inc.) to P. Le (NETL), March 27,1997.

12. NETL/Gasification Technologies team website (System Studies FY2000) : http://www.netl.doe.gov/coalpower/gasification/system/shell3x_.pdf

13. NETL/Gasification Technologies team website (Publications) : http://www.netl.doe.gov/coalpower/gasification/pubs/pdf/igcc-co2.pdf

14. D. J. White , "High Efficiency Hybrid Gas Turbine And Hydraulic Air Compressor – Final Report", TRITEX Consulting , Final Report Number TRC-101-2001.

15. Private Communication, E. Parsons (NETL) to W. Shelton (EG&G), October 10, 2002.

16. D. E. Fain (ORNL), "Research and Development on Hydrogen Separation Technology with Inorganic Membranes", Pittsburgh Coal Conference 1999.

17. T.L. Buchanan, M. G. Klett, M. D. Rutkowski and J.S. White, "Decarbonized Fuel Production Facilities Base Case Comparisons ", Letter Report, June 1999, Prepared For: The United States Department of Energy, Office of Fossil Energy, Contract No. DE-AMO1-98FE65271 BY: Parsons Infrastructure & Technology, Reading, Pennsylvania.

18. Private Communication , Patterson Pump Company (Toccoa, Georgia) to J. Lyons (EG&G), October 20, 2002.

19. Clean Energy Systems, Inc., the proposed technology is covered by one or more of the following U.S. Patents: 5,437,899; 5,590,528; 5,680,764; 5,709,077; 5,715,673; 5,956,937; 5,970,702; 6,170,264; 6,206,684; 6,247,316. Additional patents are pending.

20. Private Communication - Larry Hoffman (Clean Energy Systems, Inc.) to NETL, July 21, 2002.

21. R. E. Anderson, H. Brandt and F. Viteri, (Clean Energy Systems, Inc.), "Power Generation from Coal with Zero Atmospheric Emissions", Eighteenth Annual International Pittsburgh Coal Conference December 3-7, 2001 Newcastle, Australia.

22. Costs were provided to NETL in a series of reports / presentations by CES to the NETL and other government organizations. (e.g. documentation from R. Smith (LLNL) to A. Layne (NETL) on proposed ZEST Research Facility, Nov. 22, 2000.)

23. J. D. Fleshman (Foster Wheeler USA), "FW Hydrogen Production", in Meyers Handbook of Petroleum Refining Processes – Second Edition.

24. The Objectives and Goals of the Vision 21 program of the U.S. DOE can be explored on http://www.netl.doe.gov/coalpower/vision21/index.html.

25. G. Cau, D. Cocco, A. Montisci, "Performance of Zero Emissions Integrated Gasification Hydrogen Combustion (ZE-IGHC) Power Plants with CO_2 Removal", ASME Turbo Expo 2001, June 2001, New Orleans, LA.

26. "Fuel Cell Handbook – Fifth Edition" , U.S. DOE NETL, October 2000.

27. "Pressurized Solid Oxide Fuel Cycle/Gas Turbine Power System" by Siemens Westinghouse / Rolls-Royce Allison for the DOE. (DE-AC26-98FT40355 , February 2000).

28. A. Rao, G. Samuelsen, (University of California – Irvine), F. Robson (kraftWork Systems, Inc.) and R. Geisbrecht (NETL), "Power Plant System Configurations for the 21ST Century.", ASME Turbo Expo 2002, June 2002, Amsterdam, Nederlands.

29. M. M. Shah , K. Mahoney, W. W. Shelton and J. Lyons, "Oxygen Transport Membranes (OTM) , CRADA (99-FO32), Task 1: OTM Integrated Gasification Combined Cycle Processes, CRADA between Praxair, Inc. and NETL , DOE, June 2000.

30. C. White (EG&G) , "SOFCELL – ASPEN Model for Solid Oxide Fuel Cell", developed for NETL – last revision 2000.

31. Private Communication – F. Robson (UTRC) and W. Day (Pratt & Whitney) to NETL, December 5, 2001.

Appendix A

Process Flow Diagrams
Material & Energy Balances

(continued)

Process Flow Diagrams
Material & Energy Balances

Pulverized Coal (PC)

PC Steam Cycle – No CO_2 Capture

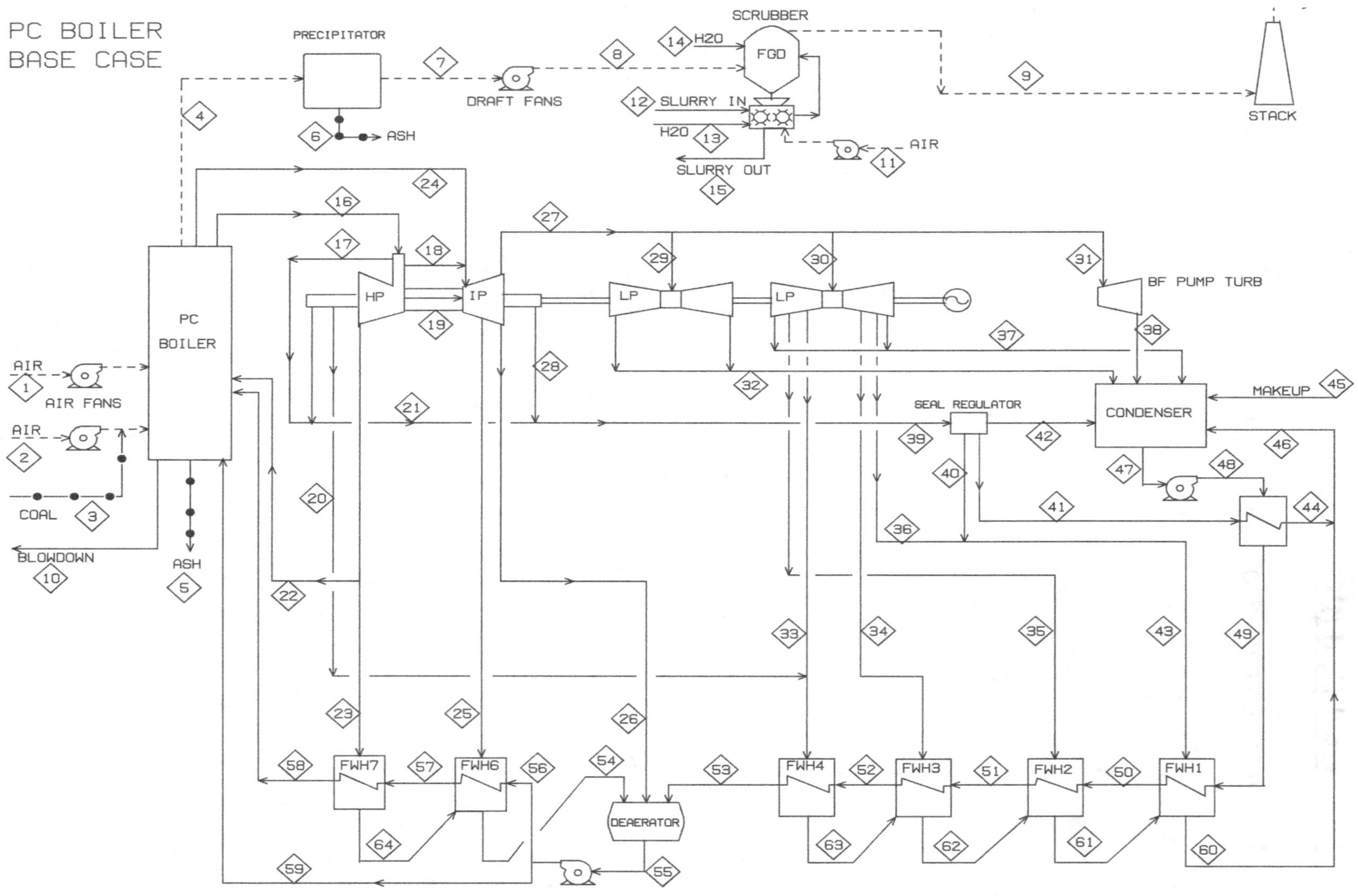
PC BOILER
BASE CASE
PRECIPITATOR
ASH
DRAFT FANS
SCRUBBER
H20
FGD
SLURRY IN
H20
SLURRY OUT
AIR
STACK
PC
BOILER
AIR
AIR FANS
AIR
COAL
BLOWDOWN
ASH
HP
IP
LP
LP
BF PUMP TURB
SEAL REGULATOR
CONDENSER
MAKEUP
FWH7
FWH6
DEAERATOR
FWH4
FWH3
FWH2
FWH1

Streams Summary

PFD ID	1	2	3	4	5	6	7
ASPEN STREAM ID	AIRFD	AIRPR	COALFEED	TOESP	ASH5	ASH6	FLUEGAS
Description	Main Air	Primary Air	Coalfeed	to ESP	Ash Boiler	Ash ESP	Fluegas
Temperature F	60	60	59	289.1	289.1	289.1	289.1
Pressure psi	14.7	14.7	14.7	14.4	14.4	14.4	14.4
Mass Flow lb/hr	2675327	821832	309464	3800348	6272	25088	3775260
Mole Flow lbmol/hr	92712	28480					127371
Enthalpy MMBtu/hr	-111.1	-34.1	-1138.9	-3696.6	-14.6	-58.5	-3638.1
Mole Frac							
O2	0.20747	0.20747		0.04557			0.04557
N2	0.77316	0.77316		0.73674			0.73674
AR	0.00921	0.00921		0.00876			0.00876
CO2	0.00030	0.00030		0.12835			0.12835
H2O	0.00986	0.00986		0.07858			0.07858
SO2				0.00190			0.00190
CL2				0.00010			0.00010
TOTAL	1.00000	1.00000		1.00000			1.00000

FLOW DIA. ID	8	9	10	11	12	13	14
ASPEN STREAM ID	20	TOSTACK	BDWN	OXIDANT	LMSTONE	SH2O	H2OMX
Description	to FGD	to Stack	H2O blowdn	Air to FGD	Lmstone	H2O - FGD	H2O -FGD
Temperature F	299.9	129	674.1	60	60	68	68
Pressure psi	15.1	14.8	2600	14.7	14.7	14.7	15
Mass Flow lb/hr	3775260	3973687	13125	61971	96893	229745	107124
Mole Flow lbmol/hr	127371	137427	729	2148	4040	12753	5946
Enthalpy MMBtu/hr	-3627.8	-4630.9	-80.3	-2.6	-623.4	-1578.8	-736.2
Mole Frac							
O2	0.04557	0.04467		0.20747			
N2	0.73674	0.69491		0.77316			
AR	0.00876	0.00826		0.00921			
CO2	0.12835	0.12058		0.00030			
H2O	0.07858	0.13134	1.00000	0.00986	1.00000	1.00000	1.00000
SO2	0.00190	0.00014		0.00000			
CL2	0.00010	0.00009		0.00000			
TOTAL	1.00000	1.00000	1.00000	1.00000			

FLOW DIAGRAM ID	15	16	17	18	19	20	21
ASPEN STREAM ID	SLURRY	H1	H2	H3	H4	H5	H7
Description	Slurry exit	Steam-HP	bleed	bld to ip	bld to ip	to FWH4	to seal reg
Temperature F	129.1	1000	1000	1000	801.6	631.4	655.4
Pressure psi	14.8	2415	2415	2415	1207.5	603.6	603.6
Mass Flow lb/hr	297308	2734080	1083	3788	32207	5521	10989
Mole Flow lbmol/hr	14719	151763	60	210	1788	306	610
Enthalpy MMBtu/hr	-1949.1	-14780.2	-5.9	-20.5	-176.7	-30.7	-60.9
Mole Frac							
H2O	1.00000	1.00000	1.00000	1.00000	1.00000	1.00000	1.00000
TOTAL	1.00000	1.00000		1.00000	1.00000		

FLOW DIAGRAM ID	22	23	24	25	26	27	28
ASPEN STREAM ID	H8	H9	H8A	I2	I3	I4	I5
Description	stm->reheat	to FWH7	Reheat->IP	to FWH6	to Deaerator	to LP Turb	to seal reg
Temperature F	631.4	631.4	1000	811.8	695.1	695.1	695.1
Pressure psi	603.6	603.6	545.4	278.9	174.9	174.9	174.9
Mass Flow lb/hr	2425661	255913	2425661	81934	160845	2215094	3784
Mole Flow lbmol/hr	134643	14205	134643	4548	8928	122955	210
Enthalpy MMBtu/hr	-13477.7	-1421.9	-12970.8	-445.6	-883.6	-12168.4	-20.8
Mole Frac							
H2O	1.00000	1.00000	1.00000	1.00000	1.00000	1.00000	1.00000
TOTAL							

FLOW DIAGRAM ID	29	30	31	32	33	34	35
ASPEN STREAM ID	L1	L2	L3	L4	L5	L6	L7
Description	to LP #1	to LP #2	to LP #3	from LP #1	to FWH4	to FWH3	to FWH2
Temperature F	695.1	695.1	695.1	101.7	479.5	293.2	205.1
Pressure psi	174.9	174.9	174.9	1	66.5	24.2	12.8
Mass Flow lb/hr	866582	1248002	100510	866582	125975	68312	64645
Mole Flow lbmol/hr	48102	69274	5579	48102	6993	3792	3588
Enthalpy MMBtu/hr	-4760.5	-6855.8	-552.1	-5070.8	-704.7	-388	-370
Mole Frac							
H2O	1.00000	1.00000	1.00000	1.00000	1.00000	1.00000	1.00000
TOTAL	1.00000	1.00000	1.00000	1.00000	1.00000	1.00000	1.00000

FLOW DIAGRAM ID	36	37	38	39	40	41	42
ASPEN STREAM ID	L8	L9	L10	S1	S2	S3	S4
Description	to FWH1	From LP #2	from LP #3	to seal reg	to FWH1	to cd reheat	to Deaer
Temperature F	172.2	110.7	113.3	625.2	625.2	625.2	625.2
Pressure psi	6.3	1.3	1.4	174.9	174.9	174.9	174.9
Mass Flow lb/hr	118188	835339	100510	14773	9545	2815	2413
Mole Flow lbmol/hr	6560	46368	5579	820	530	156	134
Enthalpy MMBtu/hr	-677.4	-4862.6	-586.5	-81.7	-52.8	-15.6	-13.3
Mole Frac							
H2O	1.00000	1.00000	1.00000	1.00000	1.00000	1.00000	1.00000
TOTAL	1.00000	1.00000	1.00000	1.00000	1.00000	1.00000	1.00000

FLOW DIAGRAM ID	43	44	45	46	47	48	49
ASPEN STREAM ID	S5	S6	MK1	C0	CD0	CDA	CD1
Description	to FWH1	to Deaer	makeup	to Deaer	from Cond	pump cdn	cdn-->FWH1
Temperature F	204.7	113.3	60	106.3	96.4	96.7	98.3
Pressure psi	6.3	1.4	14.7	1.4	0.9	330	321
Mass Flow lb/hr	127733	2815	13125	395001	2248513	2248513	2248513
Mole Flow lbmol/hr	7090	156	729	21926	124810	124810	124810
Enthalpy MMBtu/hr	-730.1	-19.1	-89.7	-2682.8	-15293.5	-15290.9	-15287.3
Mole Frac							
H2O	1.00000	1.00000	1.00000	1.00000	1.00000	1.00000	1.00000
TOTAL	1.00000	1.00000	1.00000	1.00000	1.00000	1.00000	1.00000

FLOW DIAGRAM ID	50	51	52	53	54	55	56
ASPEN STREAM ID	CD2	CD3	CD4	CD5	C76	C5	P1
Description	cdn-->FWH2	cdn-->FWH3	cdn-->FWH4	to Deaer	to Deaer	from Deaer	to FWH6
Temperature F	167.8	199.4	232	293.5	405.7	365.9	372.3
Pressure psi	300	250	210	175	263.8	164.8	2903.3
Mass Flow lb/hr	2248513	2248513	2248513	2248513	337847	2747205	2652976
Mole Flow lbmol/hr	124810	124810	124810	124810	18753	152491	147261
Enthalpy MMBtu/hr	-15131.7	-15060.8	-14987.3	-14847.2	-2190.4	-17933.2	-17289.7
Mole Frac							
H2O	1.00000	1.00000	1.00000	1.00000	1.00000	1.00000	1.00000
TOTAL	1.00000	1.00000	1.00000	1.00000	1.00000	1.00000	1.00000

FLOW DIAGRAM ID	57	58	59	60	61	62	63	64
ASPEN STREAM ID	P3	P4	P2	C1	C2	C3	C4	C7
Description	to FWH7	to econ	to pc	from FWH1	from FWH2	from FWH3	from FWH4	from FWH7
Temperature F	404.3	485.5	372.3	106.2	175.4	206.7	239.1	415.5
Pressure psi	2620	2758	2903.3	6	11.9	22.4	62.4	588.5
Mass Flow lb/hr	2652976	2652976	94229	392186	264453	199808	131495	255913
Mole Flow lbmol/hr	147261	147261	5230	21769	14679	11091	7299	14205
Enthalpy MMBtu/hr	-17201.6	-16966.7	-614.1	-2663.7	-1777.8	-1337	-875.6	-1656.8
Mole Frac								
H2O	1.00000	1.00000	1.00000	1.00000	1.00000	1.00000	1.00000	1.00000
TOTAL	1.00000	1.00000	1.00000	1.00000	1.00000	1.00000	1.00000	1.00000

POWER SUMMARY PC Steam Cycle - No CO2 Capture

TURBINE SECTION	POWER KW
HP TURBINE	-119608.27
IP TURBINE	-102672.41
LP TURBINE #1	-90957.98
LP TURBINE #2	-113113.83
TOTAL TURBINE	**-426352.49**
GENERATOR LOSS	6395.29
NET STEAM TURBINE	**-419957.19**

	POWER KW	
LP Turb #3	-10068.36	(lp turb #3 supplies power for
HP Pump	8632.28	the hp feedwater pump)
extra	-1436.08	

***NOTE - ASPEN sign convention
   "-" power produced
   "+" power required

DRAFT FANS	POWER KW
- Primary Air	915.19
- Forced	871.17
- Induced	3057.95
TOTAL FANS	**4844.31**

MISC WORK	17564.86
CONDENSER PUMP	755.99

NET POWER (MWe)	-396.79
COALFEED (LBS/HR)	309464.00
EFF % (HHV)	37.51

Pulverized Coal (PC)

PC Steam Cycle - Amine CO_2 Capture

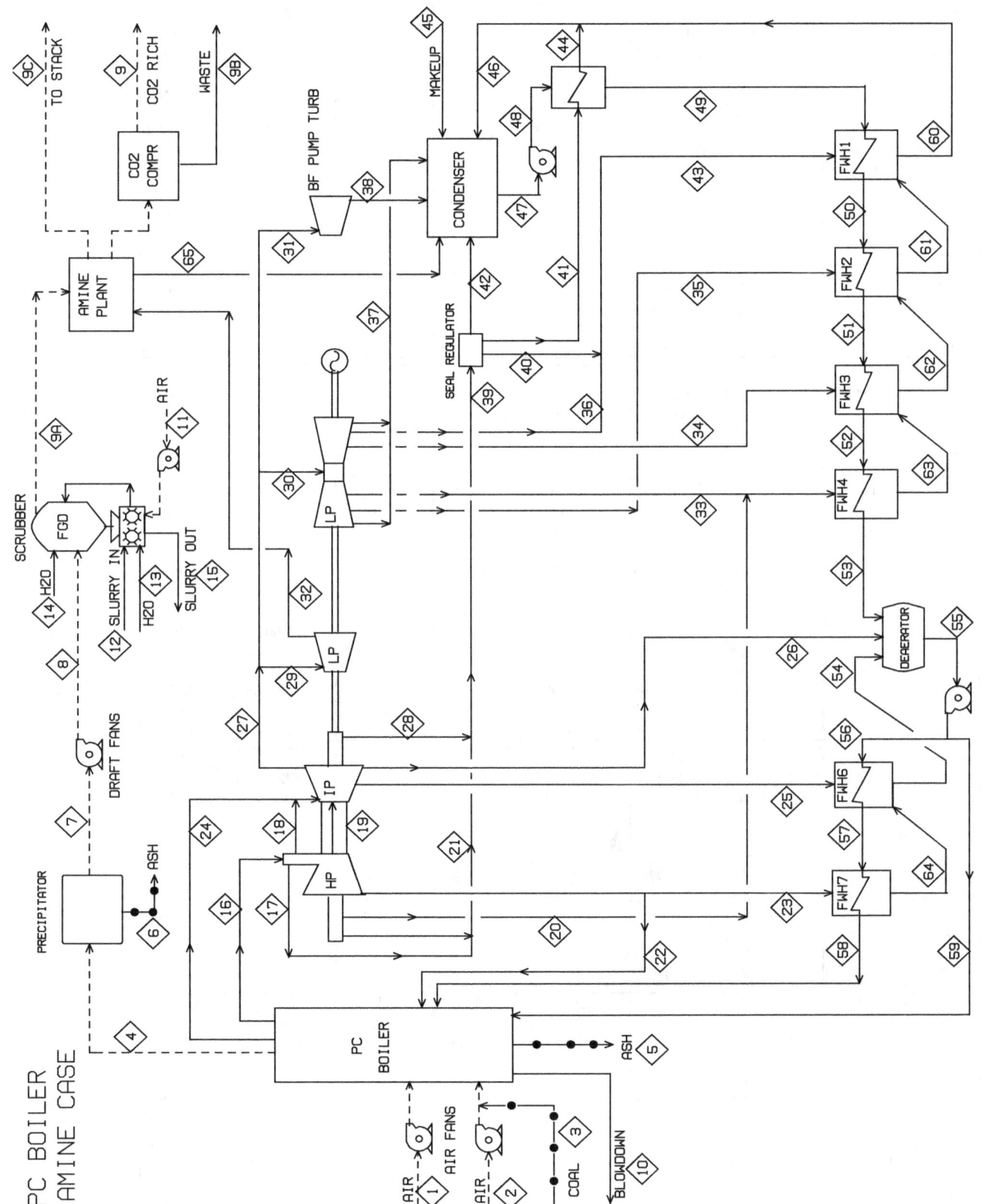

PC BOILER
AMINE CASE

A-9

FLOW DIA. ID	1	2	3	4	5	6	7
ASPEN STREAM ID	AIRFD	AIRPR	COALFEED	TOESP	ASH5	ASH6	FLUEGAS
Description	Main Air	Primary Air	Coalfeed	to ESP	Ash Boiler	Ash ESP	Fluegas
Temperature F	60	60	59	289.1	289.1	289.1	289.1
Pressure psi	14.7	14.7	14.7	14.4	14.4	14.4	14.4
Mass Flow lb/hr	2675327	821832	309464	3800348	6272	25088	3775260
Mole Flow lbmol/hr	92712	28480					127371
Enthalpy MMBtu/hr	-111.1	-34.1	-1138.9	-3696.6	-14.6	-58.5	-3638.1
Mole Frac							
O2	0.20747	0.20747		0.04557			0.04557
N2	0.77316	0.77316		0.73674			0.73674
AR	0.00921	0.00921		0.00876			0.00876
CO2	0.00030	0.00030		0.12835			0.12835
H2O	0.00986	0.00986		0.07858			0.07858
SO2	0.00000	0.00000		0.00190			0.00190
CL2	0.00000	0.00000		0.00010			0.00010
TOTAL	1.00000	1.00000		1.00000			1.00000

FLOW DIA. ID	8	9A	9B	9C	9	11	12
ASPEN STREAM ID	20	TOAMINE	LIQW	STACKGAS	CO2HP	OXIDANT	LMSTONE
Description	to FGD	to MEA	liquid waste	to stack	HP CO2	Air to FGD	Lmstone
Temperature F	299.9	129	95.7	101	228	60	60
Pressure psi	15.1	14.8	14.7	14.7	1500	14.7	14.7
Mass Flow lb/hr	3775260	3973687	234034	3058207	692907	61971	96893
Mole Flow lbmol/hr	127371	137427	12979	110132	15772	2148	4040
Enthalpy MMBtu/hr	-3627.8	-4630.9	-1603.4	-1040.2	-2665.1	-2.6	-623.4
Mole Frac							
O2	0.04557	0.04467	0.00000	0.05575	0.00006	0.20747	0.00000
N2	0.73674	0.69491	0.00000	0.86727	0.00045	0.77316	0.00000
AR	0.00876	0.00826	0.00000	0.01031	0.00000	0.00921	0.00000
CO2	0.12835	0.12058	0.00044	0.00741	0.99677	0.00030	0.00000
H2O	0.07858	0.13134	0.99951	0.05910	0.00272	0.00986	1.00000
SO2	0.00190	0.00014	0.00000	0.00000	0.00000	0.00000	0.00000
CL2	0.00010	0.00009	0.00000	0.00000	0.00000	0.00000	0.00000
MEA	0.00000	0.00000	0.00005	0.00016	0.00000	0.00000	0.00000
TOTAL	1.00000	1.00000	1.00000	1.00000	1.00000	1.00000	1.00000

FLOW DIAGRAM ID	10	13	14	15	16	17	18
ASPEN STREAM ID	BDWN	SH2O	H2OMX	SLURRY	H1	H2	H3
Description	H2O - bldn	H2O - FGD	H2O -FGD	Slurry exit	Steam-HP	bleed	bld to ip
Temperature F	674.1	68	68	129.1	1000	1000	1000
Pressure psi	2600	14.7	15	14.8	2415	2415	2415
Mass Flow lb/hr	13125	229745	107124	297308	2734080	1083	3788
Mole Flow lbmol/hr	729	12753	5946	14719	151763	60	210
Enthalpy MMBtu/hr	-80.3	-1578.8	-736.2	-1949.1	-14780.2	-5.9	-20.5
Mole Frac							
H2O	1.00000	1	1	1	1	1	1
TOTAL	1.00000	1.00000	1.00000	1.00000	1.00000	1.00000	1.00000

FLOW DIAGRAM ID	19	20	21	22	23	24	25
ASPEN STREAM ID	H4	H5	H7	H8	H9	H8A	I2
Description	bld to ip	to FWH4	to seal reg	stm->reheat	to FWH7	Reheat->IP	to FWH6
Temperature F	801.6	631.4	655.4	631.4	631.4	1000	811.8
Pressure psi	1207.5	603.6	603.6	603.6	603.6	545.4	278.9
Mass Flow lb/hr	32207	5521	10989	2425661	255913	2425661	81934
Mole Flow lbmol/hr	-176.7	-30.7	-60.9	-13477.7	-1421.9	-12970.8	-445.6
Enthalpy MMBtu/hr	1788	306	610	134643	14205	134643	4548
Mole Frac							
H2O	1	1	1	1	1	1	1
TOTAL	1.00000	1.00000	1.00000	1.00000	1.00000	1.00000	1.00000

FLOW DIAGRAM ID	26	27	28	29	30	31	32
ASPEN STREAM ID	I3	I4	I5	L1	L2	L3	STMAMN
Description	to Deaerator	to LP Turb	to seal reg	to LP #1	to LP #2	to LP #3	to MEA
Temperature F	695.1	695.1	695.1	695.1	695.1	695.1	372.4
Pressure psi	174.9	174.9	174.9	174.9	174.9	174.9	35
Mass Flow lb/hr	160845	2215094	3784	1276467	838117	100510	1276467
Mole Flow lbmol/hr	8928	122955	210	70854	46522	5579	70854
Enthalpy MMBtu/hr	-883.6	-12168.4	-20.8	-7012.1	-4604.1	-552.1	-7203
Mole Frac							
H2O	1	1	1	1	1	1	1
TOTAL	1.00000	1.00000	1.00000	1.00000	1.00000	1.00000	1.00000

FLOW DIAGRAM ID	33	34	35	36	37	38	39
ASPEN STREAM ID	L5	L6	L7	L8	L9	L10	S1
Description	to FWH4	to FWH3	to FWH2	to FWH1	from LP #2	from LP #3	to seal reg
Temperature F	479.5	293.2	205.1	172.2	110.7	113.3	625.2
Pressure psi	66.5	24.2	12.8	6.3	1.3	1.4	174.9
Mass Flow lb/hr	125975	24300	27946	40100	596082	100510	14773
Mole Flow lbmol/hr	6993	1349	1551	2226	33087	5579	820
Enthalpy MMBtu/hr	-704.7	-138	-159.9	-229.8	-3469.9	-586.5	-81.7
Mole Frac							
H2O	1	1	1	1	1	1	1
TOTAL	1.00000	1.00000	1.00000	1.00000	1.00000	1.00000	1.00000

FLOW DIAGRAM ID	40	41	42	43	44	45	46
ASPEN STREAM ID	S2	S3	S4	S5	S6	MK1	C0
Description	to FWH1	to cd reheat	to Deaer	to FWH1	to Deaer	makeup	to Deaer
Temperature F	625.2	625.2	625.2	255.9	113.3	60	106.3
Pressure psi	174.9	174.9	174.9	6.3	1.4	14.7	1.4
Mass Flow lb/hr	9545	1408	3820	49645	1408	13125	234795
Mole Flow lbmol/hr	530	78	212	2756	78	729	13033
Enthalpy MMBtu/hr	-52.8	-7.8	-21.1	-282.6	-9.5	-89.7	-1594.7
Mole Frac							
H2O	1	1	1	1	1	1	1
TOTAL	1.00000	1.00000	1.00000	1.00000	1.00000	1.00000	1.00000

FLOW DIAGRAM ID	47	48	49	50	51	52	53
ASPEN STREAM ID	CD0	CDA	CD1	CD2	CD3	CD4	CD5
Description	from Cond	Pump cdn	cdn-->FWH1	cdn-->FWH2	cdn-->FWH3	cdn-->FWH4	to Deaer
Temperature F	96.4	96.7	98.6	168.1	202	231.6	293.5
Pressure psi	0.9	330	321	300	250	210	175
Mass Flow lb/hr	972046	972046	972046	972046	972046	972046	2248513
Mole Flow lbmol/hr	53956	53956	53956	53956	53956	53956	124810
Enthalpy MMBtu/hr	-6611.5	-6610.3	-6608.5	-6541.2	-6508.4	-6479.5	-14847.2
Mole Frac							
H2O	1	1	1	1	1	1	1
TOTAL	1.00000	1.00000	1.00000	1.00000	1.00000	1.00000	1.00000

FLOW DIAGRAM ID	54	55	56	57	58	59	60
ASPEN STREAM ID	C76	C5	P1	P3	P4	P2	C1
Description	to Deaer	from Deaer	to FWH6	to FWH7	to econ	to pc	from FWH1
Temperature F	405.7	365.9	372.3	404.3	485.5	372.3	106.2
Pressure psi	263.8	164.8	2903.3	2620	2758	2903.3	6
Mass Flow lb/hr	337847	2747205	2652976	2652976	2652976	94229	233387
Mole Flow lbmol/hr	18753	152491	147261	147261	147261	5230	12955
Enthalpy MMBtu/hr	-2190.4	-17933.2	-17289.7	-17201.6	-16966.7	-614.1	-1585.1
Mole Frac							
H2O	1	1	1	1	1	1	1
TOTAL	1.00000	1.00000	1.00000	1.00000	1.00000	1.00000	1.00000

Amine Case - Stream Summary

FLOW DIAGRAM ID	61	62	63	64	65		
ASPEN STREAM ID	C2	C3	C4	C7	11		
Description	from FWH2	from FWH3	from FWH4	from FWH7	from MEA		
Temperature F	175.4	206.7	239.1	415.5	232.3		
Pressure psi	11.9	22.4	62.4	588.5	215		
Mass Flow lb/hr	183742	155796	131496	255913	1276467		
Mole Flow lbmol/hr	10199	8648	7299	14205	70854		
Enthalpy MMBtu/hr	-1235.2	-1042.5	-875.6	-1656.8	-8507.8		
Mole Frac							
H2O	1	1	1	1	1		
TOTAL	1.00000	1.00000	1.00000	1.00000	1.00000		

POWER SUMMARY - BASE CASE modified for providing steam to amine system reboiler
(Basis - CO_2 in exit gas = 692806 lbs/hr   - 95% of CO_2 generated,
reboiler duty in amine system = 4.08 MMBTU/Metric Ton CO_2,
Steam provided from steam cycle at 35 psia and 372 F, Condensate return at 215 psia and 232 F,
Steam flowrate = 1276467 lbs/hr)

TURBINE SECTION	POWER KW	
HP TURBINE	-119608.27	
IP TURBINE	-102672.41	
LP TURBINE #1	-55936.54	****
LP TURBINE #2	-75546.58	****
TOTAL TURBINE	-353763.80	
GENERATOR LOSS	-5306.457	
NET STEAM TURBINE	-348457.34	

	POWER KW	
LP Turb #3	-10068.36	(lp turb #3 supplies power for
HP Pump	8632.28	the hp feedwater pump)
extra	-1436.08	

***NOTE - ASPEN sign convention
    "-" power produced
    "+" power required

**** POWER REDUCED FROM BASE CASE DUE TO STEAM
EXTRACTION FOR AMINE SYSTEM REBOILER,
LP Turbine section #2 was modified by reducing bleeds, assumes
returning steam sent to the reboiler as condensate at 215 psia and 232 F

DRAFT FANS	POWER KW
- Primary Air	915.19
- Forced	871.17
- Induced	3057.95
TOTAL FANS	4844.31

MISC WORK	17564.86	
CONDENSER PUMP	343.09	(REDUCED SINCE CONDSEROR FLOW IS REDUCED
COND. RETURN- AMINE	266.47	(PUMPS REBOILER CONDENSATE FROM 25 PSIA TO 215 PSIA)
Amine plant	12567.90	(calculated as 40 kWh/metric ton co2 * 692806/2205 metric ton co2)
CO2 COMPRESSOR	29791.38	

717334.895 = compr inlet lb/hr
709099.756 = compr outlet lb/hr   (692806 lb/hr CO_2)

NET POWER (MWe)	-283.08
COALFEED (LBS/HR)	309464.00
EFF % (HHV)	26.76

Pulverized Coal (PC)

PC Steam Cycle - O_2 Boiler / CO_2 Capture

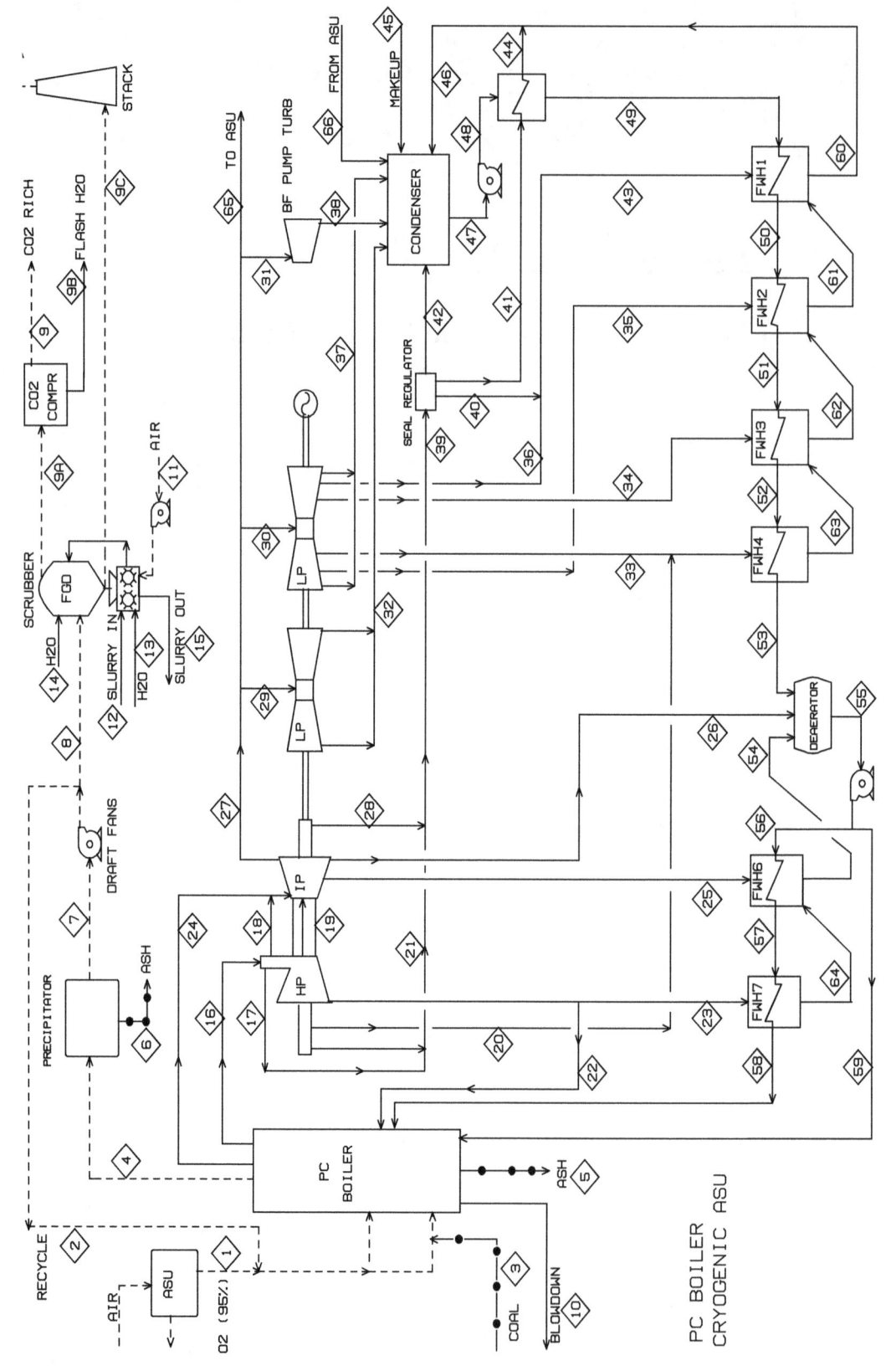

PC BOILER
CRYOGENIC ASU

A-14

PC Steam Cycle – O2 Boiler/CO2 Capture - Stream Summary

FLOW DIA. ID	1	2	3	4	5	6	7
ASPEN STREAM ID	O2CRYO	RCYIN	COALFEED	TOESP	ASH5	ASH6	FLUEGAS
Description	O2 (95%)	RECYCLE	Coalfeed coal	to ESP	Ash Boiler solids	Ash ESP Solids	Fluegas
Temperature F	60	305	59	306	306	306	306
Pressure psi	18	15.1	14.7	14.4	14.4	14.4	14.4
Mass Flow lb/hr	668508	2400581	296097	3359182	6001	24004	3335178
Mole Flow lbmol/hr	20750	68484					95146
Enthalpy MMBtu/hr	-2.6	-8972.6	-1089.7	-12520.7	-14	-55.8	-12464.9
Mole Frac							
O2	0.95000	0.04534		0.04534			0.04534
N2	0.01500	0.01664		0.01664			0.01664
AR	0.03500	0.02725		0.02724			0.02724
CO2		0.58536		0.58536			0.58536
H2O		0.31627		0.31628			0.31628
SO2		0.00868		0.00868			0.00868
CL2		0.00046		0.00046			0.00046
TOTAL	1.00000	1.00000		1.00000			1.00000

FLOW DIA. ID	8	9A	9B	9C	9	11	15
ASPEN STREAM ID	21	FLVAP1	H2OWST	TOSTACK	37	OXIDANT	LIQWST
Description	to FGD	to Flash	H2O-Flash	to Stack	CO2 Prod	Oxid to FGD	Slurry exit 12.2% solids
Temperature F	316.9	129	83.8	129	231	60	129
Pressure psi	15.3	14.7	14.7	14.7	1500	14.7	14.7
Mass Flow lb/hr	934610	828534	49455	60821	779080	59291	299440
Mole Flow lbmol/hr	26663	21012	2745	2222	18267	2055	14910
Enthalpy MMBtu/hr	-3490.3	-2958.9	-339	-29.9	-2677.6	-2.5	-1966
Mole Frac							
O2	0.04534	0.05753	0.00000	0.14394	0.06617	0.20747	7.1496E-07
N2	0.01664	0.02112	0.00000	0.71498	0.02429	0.77316	2.3719E-07
AR	0.02724	0.03457	0.00000	0.00851	0.03976	0.00921	4.0633E-08
CO2	0.58536	0.75289	0.00003	0.00038	0.86603	0.00030	7.7199E-09
H2O	0.31628	0.13243	0.99996	0.13218	0.00206	0.00986	0.999999
SO2	0.00868	0.00088	0.00000	0.00001	0.00101	0.00000	1.3246E-08
CL2	0.00046	0.00058	0.00000	0.00000	0.00067	0.00000	1.3052E-09
TOTAL	1.00000	1.00000	1.00000	1.00000	1.00000	1.00000	1.00000

FLOW DIAGRAM ID	10	12	13	14	16	17	18
ASPEN STREAM ID	BDWN	LMSTONE	SH2O	H2OMX	H1	H2	H3
description	H2O blowdn	Lmstone 30% solids	H2O - Slurry	H2O - FGD	Steam-HP	Bleed	bld to ip
Temperature F	674.1	60	68	68	1000	1000	1000
Pressure psi	2600	14.7	14.7	14.7	2415	2415	2415
Mass Flow lb/hr	13119	92708	69692	32495	2732657	1082	3786
Mole Flow lbmol/hr	728	3861	3868	1804	151684	60	210
Enthalpy MMBtu/hr	-80.3	-595.6	-478.9	-223.3	-14772.5	-5.9	-20.5
Mole Frac							
H2O	1.00000	1.00000	1.00000	1.00000	1.00000	1.00000	1.00000
TOTAL	1.00000	1.00000	1.00000	1.00000	1.00000	1.00000	1.00000

FLOW DIAGRAM ID	19	20	21	22	23	24	25
ASPEN STREAM ID	H4	H5	H7	H8	H9	H8A	I2
description	bld to ip	to FWH4	to seal reg	stm->reheat	to FWH7	Reheat->IP	to FWH6
Temperature F	801.6	631.4	655.4	631.4	631.4	1000	811.8
Pressure psi	1207.5	603.6	603.6	603.6	603.6	545.4	278.9
Mass Flow lb/hr	32191	5518	10983	2424399	255780	2424399	81891
Mole Flow lbmol/hr	1787	306	610	134573	14198	134573	4546
Enthalpy MMBtu/hr	-176.6	-30.7	-60.9	-13470.7	-1421.2	-12964.1	-445.3
Mole Frac							
H2O	1.00000	1.00000	1.00000	1.00000	1.00000	1.00000	1.00000
TOTAL	1.00000	1.00000	1.00000	1.00000	1.00000	1.00000	1.00000

FLOW DIAGRAM ID	26	27	28	29	30	31	32
ASPEN STREAM ID	I3	I4	I5	L1	L2	L3	L4
Description	to Deaerator	to LP Turb	to seal reg	to LP #1	to LP #2	to LP #3	from LP #1
Temperature F	695.1	695.1	695.1	695.1	695.1	695.1	101.7
Pressure psi	174.9	174.9	174.9	174.9	174.9	174.9	1
Mass Flow lb/hr	160762	2213941	3782	866131	1247352	100458	845882
Mole Flow lbmol/hr	8924	122891	210	48077	69238	5576	46953
Enthalpy MMBtu/hr	-883.1	-12162.1	-20.8	-4758	-6852.2	-551.9	-4949.7
Mole Frac							
H2O	1.00000	1.00000	1.00000	1.00000	1.00000	1.00000	1.00000
TOTAL	1.00000	1.00000	1.00000	1.00000	1.00000	1.00000	1.00000

FLOW DIAGRAM ID	33	34	35	36	37	38	39
ASPEN STREAM ID	L5	L6	L7	L8	L9	L10	S1
Description	to FWH4	to FWH3	to FWH2	to FWH1	from LP #2	from LP #3	to seal reg
Temperature F	479.5	293.2	205.1	172.2	110.7	113.3	625.2
Pressure psi	66.5	24.2	12.8	6.3	1.3	1.4	174.9
Mass Flow lb/hr	125909	68277	64612	118127	834904	100458	14765
Mole Flow lbmol/hr	6989	3790	3586	6557	46344	5576	820
Enthalpy MMBtu/hr	-704.4	-387.8	-369.8	-677	-4860.1	-586.2	-81.6
Mole Frac							
H2O	1.00000	1.00000	1.00000	1.00000	1.00000	1.00000	1.00000
TOTAL	1.00000	1.00000	1.00000	1.00000	1.00000	1.00000	1.00000

FLOW DIAGRAM ID	40	41	42	43	44	45	46
ASPEN STREAM ID	S2	S3	S4	S5	S6	MK1	C0
Description	to FWH1	to cd reheat	to Deaer	to FWH1	to Deaer	makeup	to Deaer
Temperature F	625.2	625.2	625.2	204.7	113.3	60	106.3
Pressure psi	174.9	174.9	174.9	6.3	1.4	14.7	1.4
Mass Flow lb/hr	9540	2814	2412	127667	2814	13119	394796
Mole Flow lbmol/hr	530	156	134	7086	156	728	21914
Enthalpy MMBtu/hr	-52.7	-15.6	-13.3	-729.8	-19.1	-89.7	-2681.4
Mole Frac							
H2O	1.00000	1.00000	1.00000	1.00000	1.00000	1.00000	1.00000
TOTAL	1.00000	1.00000	1.00000	1.00000	1.00000	1.00000	1.00000

FLOW DIAGRAM ID	47	48	49	50	51	52	53
ASPEN STREAM ID	CD0	CDA	CD1	CD2	CD3	CD4	CD5
Description	from Cond	Pump cdn	cdn-->FWH1	cdn-->FWH2	cdn-->FWH3	cdn-->FWH4	to Deaer
Temperature F	96.4	96.7	98.3	167.8	199.4	232	293.5
Pressure psi	0.9	330	321	300	250	210	175
Mass Flow lb/hr	2247343	2247343	2247343	2247343	2247343	2247343	2247343
Mole Flow lbmol/hr	124745	124745	124745	124745	124745	124745	124745
Enthalpy MMBtu/hr	-15285.5	-15282.9	-15279.4	-15123.8	-15052.9	-14979.5	-14839.4
Mole Frac							
H2O	1.00000	1.00000	1.00000	1.00000	1.00000	1.00000	1.00000
TOTAL	1.00000	1.00000	1.00000	1.00000	1.00000	1.00000	1.00000

FLOW DIAGRAM ID	54	55	56	57	58	59	60
ASPEN STREAM ID	C76	C5	P1	P3	P4	P2	C1
Description	to Deaer	From Deaer	to FWH6	to FWH7	to econ	to pc	from FWH1
Temperature F	405.7	365.9	372.3	404.3	485.5	372.3	106.2
Pressure psi	263.8	164.8	2903.3	2620	2758	2903.3	6
Mass Flow lb/hr	337671	2745776	2651596	2651596	2651596	94180	391982
Mole Flow lbmol/hr	18743	152412	147184	147184	147184	5228	21758
Enthalpy MMBtu/hr	-2189.3	-17923.9	-17280.7	-17192.7	-16957.9	-613.8	-2662.3
Mole Frac							
H2O	1.00000	1.00000	1.00000	1.00000	1.00000	1.00000	1.00000
TOTAL	1.00000	1.00000	1.00000	1.00000	1.00000	1.00000	1.00000

PC Steam Cycle – O2 Boiler/CO2 Capture - Stream Summary

FLOW DIAGRAM ID	61	62	63	64	65	66	
ASPEN STREAM ID	C2	C3	C4	C7	STMEXT	CNDSASU	
Description	from FWH2	From FWH3	from FWH4	from FWH7	to ASU	from ASU	
Temperature F	175.4	206.7	239.1	415.5	695.1	370.7	
Pressure psi	11.9	22.4	62.4	588.5	174.9	174.9	
Mass Flow lb/hr	264315	199704	131427	255780	20249	20249	
Mole Flow lbmol/hr	14672	11085	7295	14198	1124	1124	
Enthalpy MMBtu/hr	-1776.9	-1336.3	-875.1	-1656	-111.2	-132.1	
Mole Frac							
H2O	1.00000	1.00000	1.00000	1.00000	1.00000	1.00000	
TOTAL	1.00000	1.00000	1.00000	1.00000	1.00000	1.00000	

POWER SUMMARY -   CRYOGENIC ASU

TURBINE SECTION	POWER KW
HP TURBINE	-119546.04
IP TURBINE	-102618.99
LP TURBINE #1	-88785.26
LP TURBINE #2	-113054.98
TOTAL TURBINE	**-424005.27**
GENERATOR LOSS	6360.08
NET STEAM TURBINE	**-417645.18**

	POWER KW	
LP Turb #3	-10063.12	(lp turb #3 supplies power for
HP Pump	8627.79	the hp feedwater pump)
extra	-1435.33	

***NOTE - ASPEN sign convention
   "-" power produced
   "+" power required

DRAFT FANS		
- Primary	small	
- Forced	small	
- Induced	2847.99	
TOTAL FANS	**2847.99**	

CO2 COMPRESSOR	33853.73

WORK ASU	64299.99
(ESTIMATE -PRAXAIR)	

MISC WORK	17468.1567
CONDENSER PUMP	755.61

NET POWER (MWe)	-298.42	
COALFEED (LBS/HR)	296097	
EFF % (HHV)	29.48	INCLUDES CO2 COMPRESSOR
EFF % (HHV)	32.83	NO CO2 COMPRESSOR

Combined Cycle

Natural Gas Combined Cycle (NGCC) - No CO_2 Capture

Natural Gas Combined Cycle (NGCC) - No CO$_2$ Capture

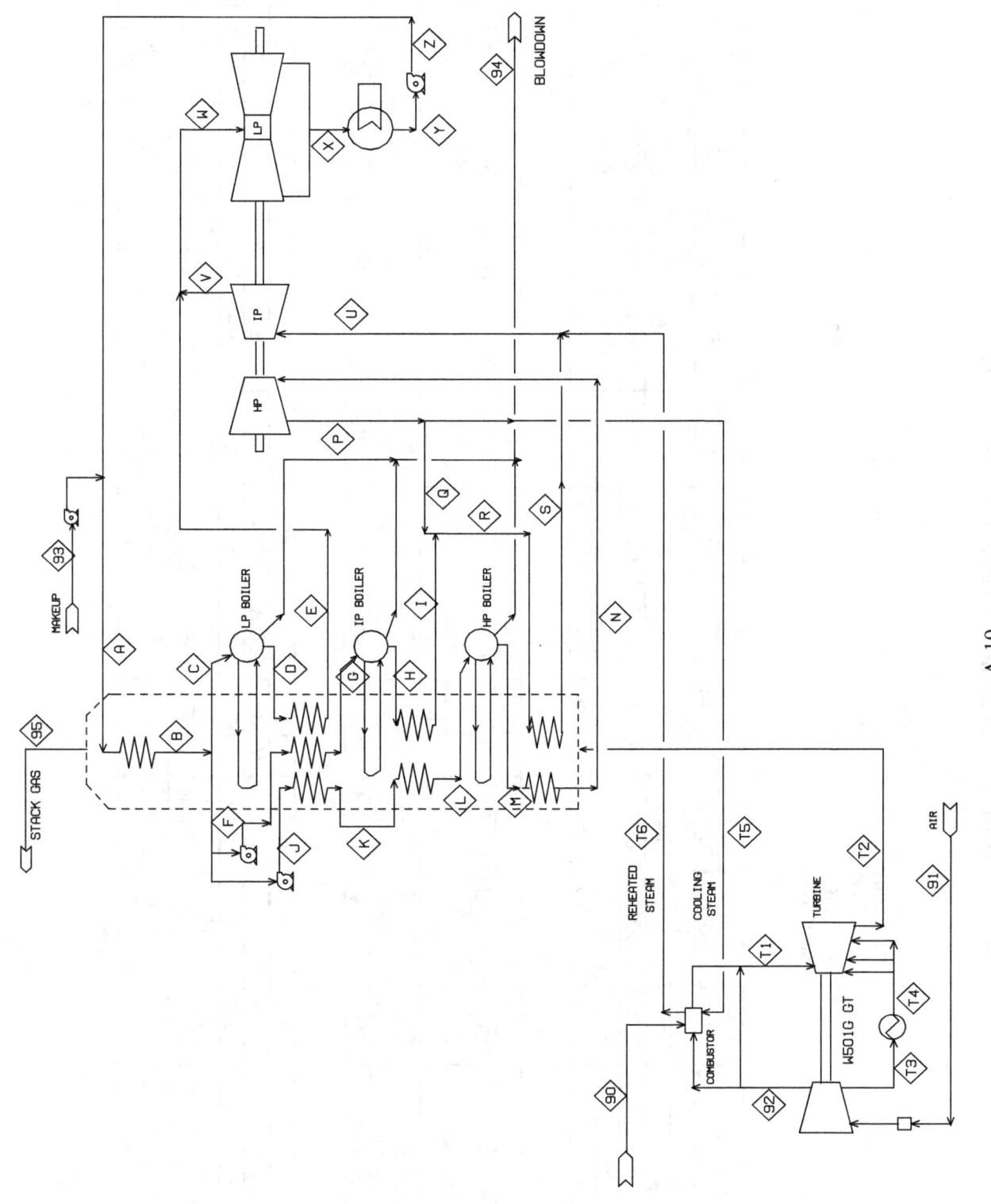

A-19

NGCC - W501G GAS TURBINE - 3 PRESSURE LEVEL STEAM CYCLE

	MWe		Efficiency;	%
Gas Turbine	266.4		LHV	57.9
Steam Turbine	121.9		HHV	52.3
Misc/Aux	9.2			
Net Power	379.1			

Stream PFD #	A	B	C	D	E	F	G	H	I	J	K	L
ASPEN Name ID	TOLPEC	HOTLP	TOLPEV	TOLPSH	LPTOIP	TOIPEC	TOIPEV	TOIPSH	FRIPSH	TOHPEC1	TOHPEC2	TOHPEV
Temperature F	92	295	295	299.3	400	296.4	463	472.8	615	299.9	463	615
Pressure psi	73.5	66.3	66.3	66.3	63	585.7	556.4	528.6	518	2263.8	2150.7	2043.1
Mass Flow lb/hr	723086	723086	86061	85201	85201	170371	170371	168667	168667	466654	466654	466654
Mole Flow lbmol/hr	40137	40137	4777	4729	4729	9457	9457	9362	9362	25903	25903	25903
Enthalpy MMBtu/hr	-4921.2	-4773.6	-568.2	-484.5	-480	-1124.3	-1094	-955	-937.7	-3076.4	-2996.3	-2907.1

Stream PFD #	M	N	P	Q	R	S	U	V	W	X	Y	Z
ASPEN Name ID	TOHPSH	TOHPTUR	FRHPTUR	TMXIP	TOREHT	52	TOIPTUR1	TOIPMX2	TOIPTUR2	TOCOND	TOCPMP	TOCMIX
Temperature F	631.5	1050	712	712	681.5	1050	1056.8	560.8	541.5	93.6	90	90.1
Pressure psi	1941	1800	518	518	518	492	492	63	63	0.8	0.7	73.5
Mass Flow lb/hr	461987	461987	461987	381987	550654	550654	630654	630654	715855	715855	715855	715855
Mole Flow lbmol/hr	25644	25644	25644	21203	30566	30566	35006	35006	39736	39736	39736	39736
Enthalpy MMBtu/hr	-2644.3	-2474	-2542.3	-2102.1	-3039.8	-2928.8	-3352	-3502.5	-3982.5	-4186	-4873.5	-4873.3

Stream PFD #	90	91	92	93	94	95	T1	T2	T3	T4	T5	T6
ASPEN Name ID	FLH2	1	2	MAKUP	TBLOW	GTPC9	31	33	3	12	C3	C4
Temperature F	200	59	813.2	80	213	208.5	2583	1100.4	813.2	600	712	1103.2
Pressure psi	400	14.7	282.2	20	15	15	268.5	15	282.2	277	518	492
Mass Flow lb/hr	103875	4467600	3933042	7231	7231	4571478	4036920	4571478	527109	527109	80000	80000
Mole Flow lbmol/hr	6475	154822	136297	401	401	161297	142772	161297	18267	18267	4441	4441
Enthalpy MMBtu/hr	-201.3	-186.6	572.4	-49.3	-45.5	-2457.2	342.6	-1367	76.7	47.9	-440.2	-423.2

Combined Cycle

Natural Gas Combined Cycle (NGCC) - CO_2 Capture

NGCC – CO2 CAPTURE

A-22

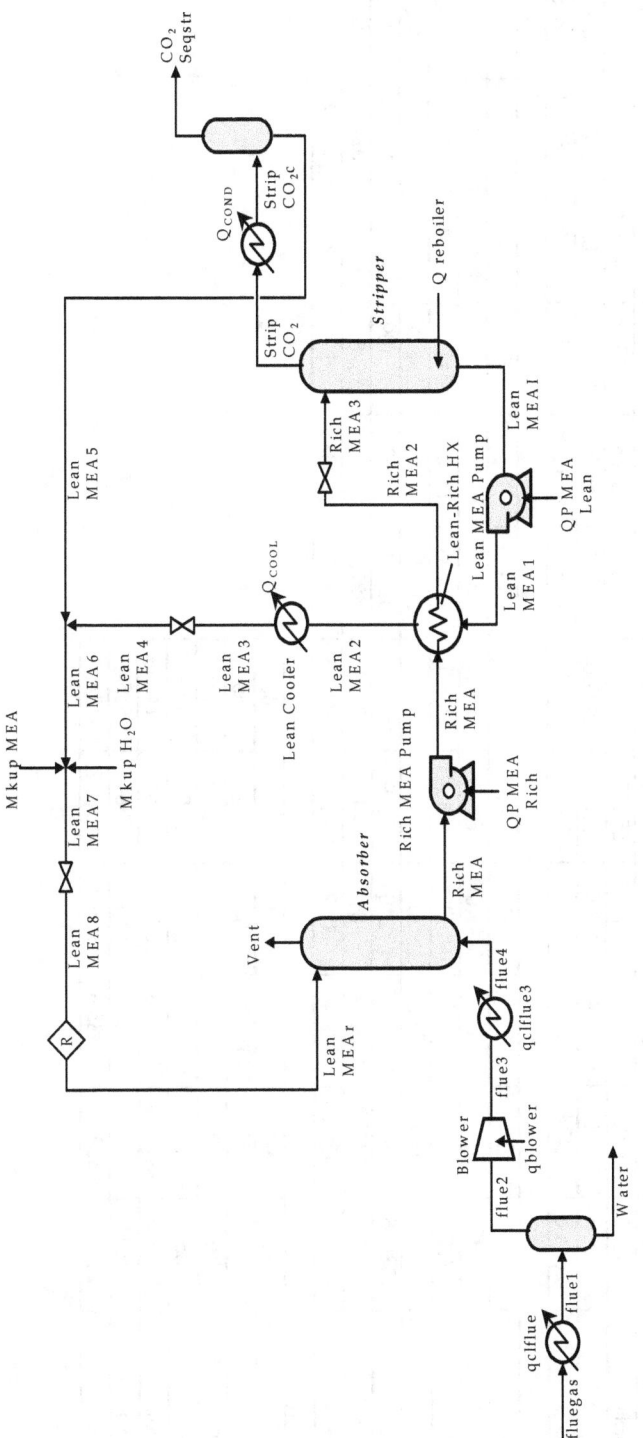

NGCC (WITH CO2 CAPTURE) - W501G GAS TURBINE - 3 PRESSURE LEVEL STEAM CYCLE

	MWe	Efficiency;	%
Gas Turbine	266.4		
Steam Turbine	90.7	LHV	49.9
Misc/Aux	30.2	HHV	45.1
Net Power	326.9		

Stream PFD #	A	B	C	D	E	F	G	H	I	J	K	L
ASPEN Name ID	TOLPEC	FRLPEC	TOLPEV	TOLPSH	LPTOIP	TOIPEC	TOIPEV	TOIPSH	FRIPSH	TOHPEC1	TOHPEC2	TOHPEV
Temperature F	90	295	295	299.3	400	296.4	463	472.8	615	299.9	463	615
Pressure psi	73.5	69.8	69.8	66.3	63	585.7	556.4	528.6	518	2263.8	2150.7	2043.1
Mass Flow lb/hr	721851	721851	84826	83978	83978	170372	170372	168668	168668	466653	466653	466653
Mole Flow lbmol/hr	40068	40068	4709	4661	4661	9457	9457	9362	9362	25903	25903	25903
Enthalpy MMBtu/hr	-4914.2	-4765.5	-560	-477.6	-473.1	-1124.3	-1094	-955	-937.7	-3076.4	-2996.3	-2907.1

Stream PFD #	M	N	P	Q	R	S	U	V	W	X	Y	Z
ASPEN Name ID	TOHPSH	TOHPTUR	FRHPTUR	TMXIP	TOREHT	52	TOIPTUR1	TOIPMX2	TOIPTUR2	TOCOND	TOCPMP	TOCMIX
Temperature F	631.5	1050	712	712	681.5	1050	1056.8	560.8	541.7	93.6	90	90.1
Pressure psi	1941	1800	518	518	518	492	492	63	63	0.8	0.7	73.5
Mass Flow lb/hr	461986	461986	461986	381986	550654	550654	630654	630654	714633	249094	714633	714633
Mole Flow lbmol/hr	25644	25644	25644	21203	30566	30566	35006	35006	39668	13827	39668	39668
Enthalpy MMBtu/hr	-2644.3	-2474	-2542.3	-2102.1	-3039.8	-2928.8	-3352	-3502.5	-3975.6	-1456.6	-4865.2	-4865

Stream PFD #	90	91	92	93	94	95	T1	T2	T3	T4	T5	T6
ASPEN Name ID	FLH2	1	2	MAKUP	TBLOW	GTPC9	31	33	3	12	C3	C4
Temperature F	200	59	813.2	80	213	208.5	2583	1100.4	813.2	600	712	1103.2
Pressure psi	400	14.7	282.2	20	15	15	268.5	15	282.2	277	518	492
Mass Flow lb/hr	103875	4467600	3933042	7219	7219	4571478	4036920	4571478	527109	527109	80000	80000
Mole Flow lbmol/hr	6475	154822	136297	401	401	161297	142772	161297	18267	18267	4441	4441
Enthalpy MMBtu/hr	-201.3	-186.6	572.4	-49.2	-45.5	-2457.3	342.6	-1367	76.7	47.9	-440.2	-423.2

Stream PFD #	40	41	42	43	44	45
ASPEN Name ID	45	53	TCPRCO2	61	62	59
Temperature F	428	250.4	140	245.2	123.1	100
Pressure psi	35	80	25.7	2100	3000	14.7
Mass Flow lb/hr	465539	465539	265986	258518	258518	114659
Mole Flow lbmol/hr	25841	25841	6296	5881	5881	6364
Enthalpy MMBtu/hr	-2614.4	-3094.5	-1033.9	-995.4	-1017.2	-784.5

Combined Cycle

IGCC Destec (E-Gas^{TM}) / CGCU / "G" Gas Turbine

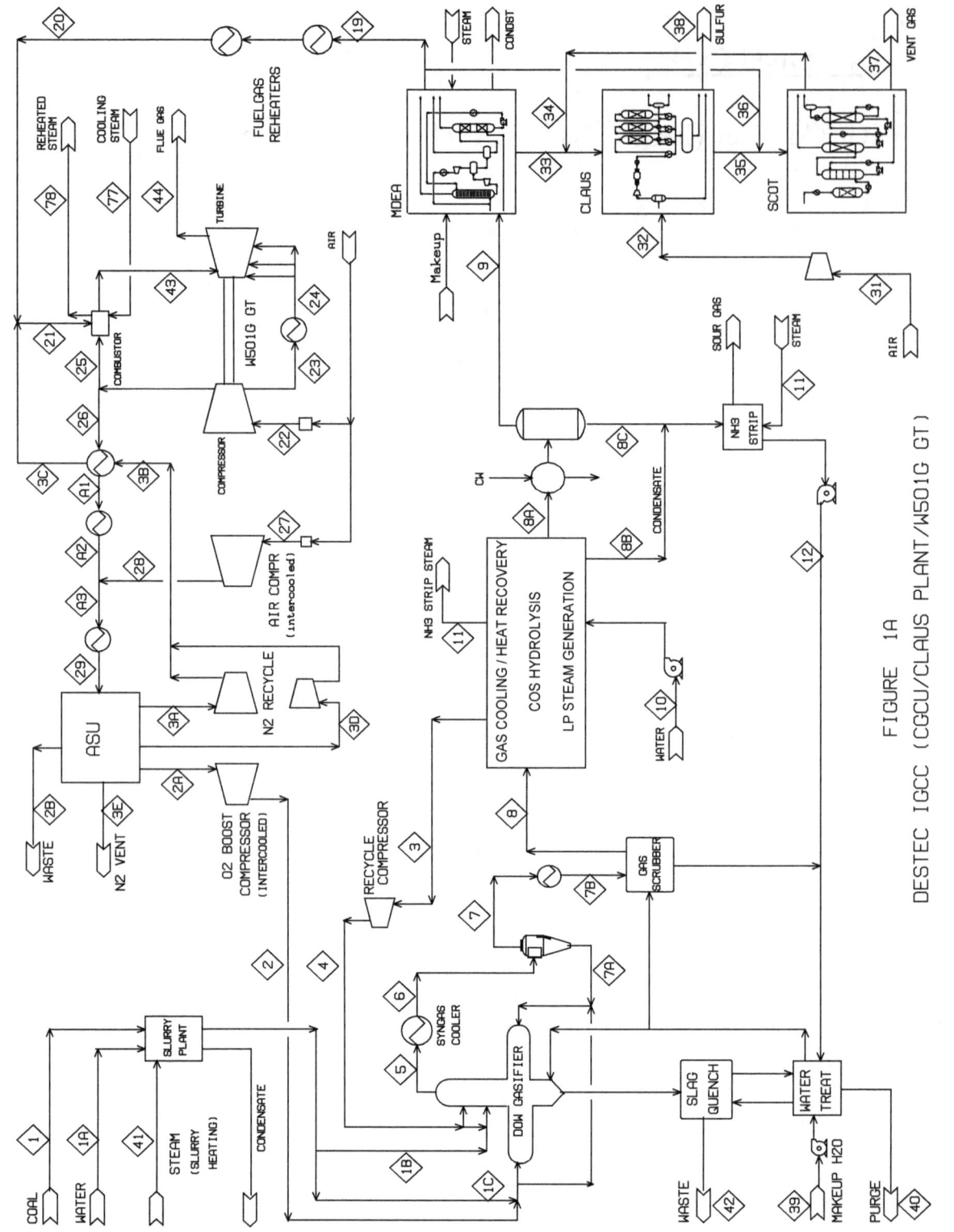

FIGURE 1A

DESTEC IGCC (CGCU/CLAUS PLANT/W501G GT)

DESTEC IGCC - (SYNGAS COOLER / CGCU / CLAUS PLANT / 3 PRES STEAM CYCLE)

SUMMARY:

POWER	MWe
GAS TURBINE	272.8
STEAM TURBINE	172.2
MISCELLANEOUS	32
AUXILIAF	12.4
NET POWER	400.6

EFFICIENCY:	%
HHV	45
LHV	46.7

STREAM	1	1A	1B	1C	2A	2B	2	3A	3B	3C	3D	3E	3	4
FLOW (LB/HR)	260226	86709	72573	274362	197846	2990	197846	317975	358735	358735	40761	270868	141102	141103
TEMPERATURE (F)	59	59	350	350	60	59	204.7	62	189.3	700	60	62	304.6	333.8
PRESSURE (PSIA)	14.7	14.7	465	465	92	14.7	472	91	300	294	265	91	378	425
H (MM BTU/HR)	-814.7	-596.7	-153.1	-574.8	-0.9	-20.3	5	-3.5	7.4	54.1	-0.3	-3	-370.7	-369.2

STREAM	5	6	7	7A	7B	8	8A	8B	8C	9	10	11	12	19
FLOW (LB/HR)	670129	670129	661340	8788	661340	705510	472085	92323	11380	460705	45000	45000	102871	424837
TEMPERATURE (F)	1900	650	649.9	649.9	415	304.2	190	232.2	101.9	103	59	280	213	116
PRESSURE (PSIA)	412	403.8	394.5	394.5	390	380	354	354	20	349	14.7	37	470	340
H (MM BTU/HR)	-1164.4	-1519.1	-1506.8	-12.4	-1568	-1853.6	-978	-617.7	-73.3	-931.2	-309.7	-255.7	-690.7	-847.5

STREAM	20	21	22	23	24	25	26	A1	A2	A3	27	28	29	31
FLOW (LB/HR)	424837	783573	4320000	527109	527109	3363310	416102	416102	416102	830440	416102	414338	830440	14107
TEMPERATURE (F)	589.7	629.2	59	813.3	600	813.3	813.3	370.4	216	210	59	203.9	190	59
PRESSURE (PSIA)	330	294	14.6	282.2	276.6	282.2	282.2	280.2	278	278	14.6	278	277	14.7
H (MM BTU/HR)	-770.3	-716.2	-180.3	76.7	47.9	489.6	60.6	13.8	-2.1	4.6	-17.4	6.7	0.5	-0.6

STREAM	32	33	34	35	36	37	38	39	40	41	42	43	44	45
FLOW (LB/HR)	14107	32346	1976	42121	6755	46900	6307	38680	87473	53025	34450	4146881	4673991	4673991
TEMPERATURE (F)	161.2	142.1	70	424	116	70	285	59	200	820.1	200	2582.2	1119.5	261
PRESSURE (PSIA)	25	18.5	17.5	26.7	340	17.5	14.7	14.7	15	150	15	268.5	15.2	14.7
H (MM BTU/HR)	-0.2	-86.3	-6.9	-109.4	-13.5	-126.8	-0.7	-266.2	-585.9	-287.8	-114.7	-265.9	-1971.2	-3032.3

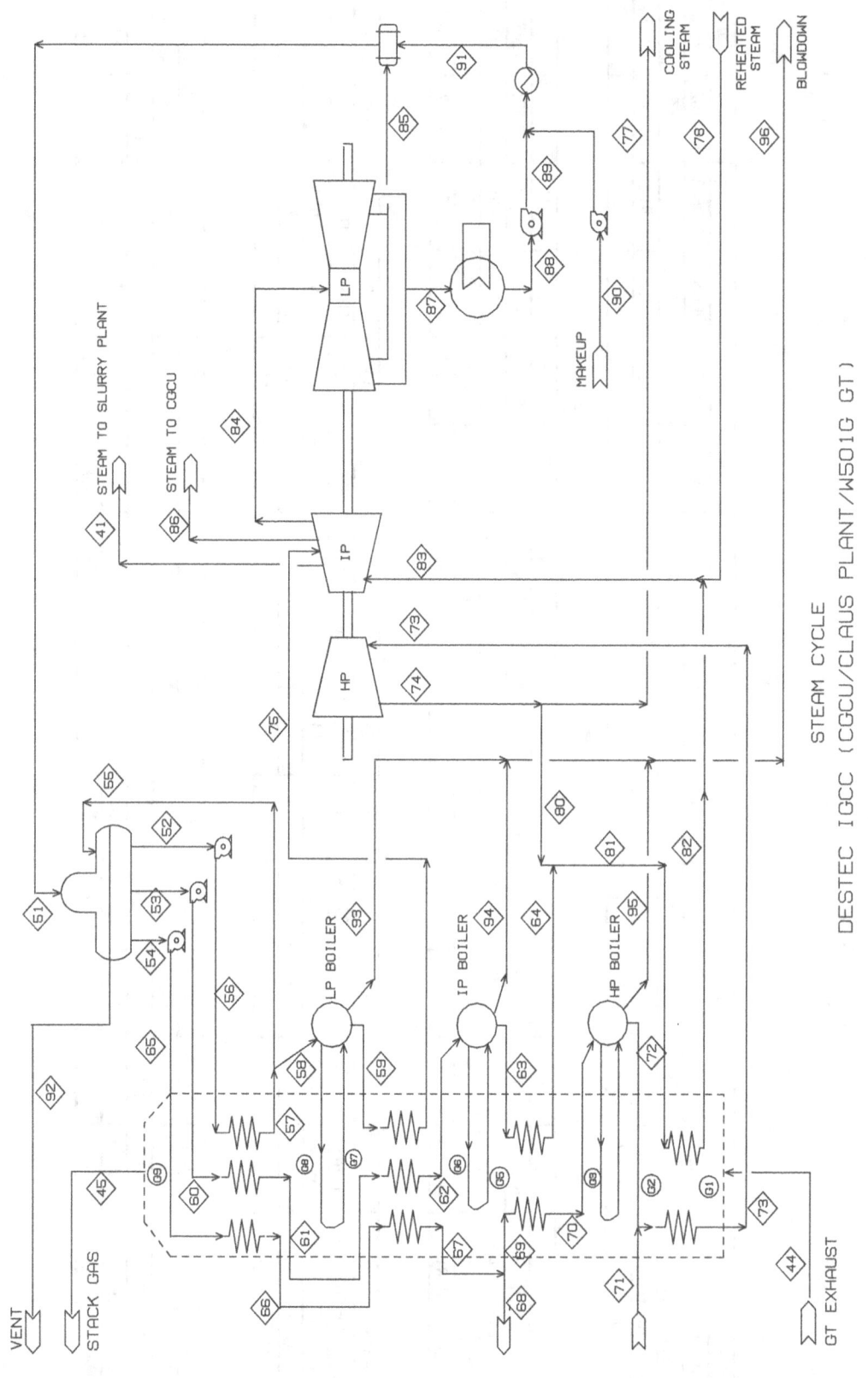

STEAM CYCLE

DESTEC IGCC (CGCU/CLAUS PLANT/W501G GT)

A-28

DESTEC IGCC - (SYNGAS COOLER / CGCU / CLAUS PLANT / 3 PRES STEAM CYCLE)

STEAM CYCLE / HRSG PROCESS STREAMS

STREAM	41	44	45	51	52	53	54	55	56	57	58	59	60	61
FLOW (LB/HR)	53025	4673991	4673991	974779	269014	259543	697343	257282	269014	269014	11732	11615	259543	259543
TEMPERATURE (F)	820.1	1119.5	261	203.8	217.3	217.3	217.3	286	217.4	286	286	305.3	218.1	286
PRESSURE (PSIA)	150	15.2	14.7	17	16.3	16.3	16.3	76.3	80.3	76.3	76.3	72.5	410.6	390
H (MM BTU/HR)	-287.8	-1971.2	-3032.3	-6525.4	-1797.2	-1733.9	-4658.8	-1700.9	-1797.1	-1778.4	-77.6	-66	-1733.5	-1715.7

STREAM	62	63	64	65	66	67	68	69	70	71	72	73	74	75
FLOW (LB/HR)	259543	256948	256948	697343	697343	697343	505841	191503	191503	505841	189588	695428	695428	11615
TEMPERATURE (F)	420	432.3	620	221.2	286	420	420	620	620	635	629.3	1050	606.7	420
PRESSURE (PSIA)	370.5	352	350	2345.6	2228.3	2116.9	2116.9	2011.1	2011.1	1910.5	1910.5	1800	350	69.5
H (MM BTU/HR)	-1679.1	-1455	-1424.5	-4652.4	-4607.2	-4510.2	-3271.6	-1191.4	-1191.4	-2888.1	-1084.7	-3724.1	-3860.5	-65.3

STREAM	77	78	80	81	82	83	84	85	86	87	88	89	90	91
FLOW (LB/HR)	70000	70000	625428	882376	882376	952376	849203	23299	61763	825904	825904	825904	125576	951480
TEMPERATURE (F)	606.7	1055.9	606.7	610.6	1050	1050.4	482	350	596.5	88.8	87.9	87.9	80	178.3
PRESSURE (PSIA)	350	342	350	350	342	342	35	17	60	0.7	0.7	40	14.7	17
H (MM BTU/HR)	-388.6	-371.8	-3472	-4896.4	-4689.6	-5061.4	-4746.9	-131.7	-341.9	-4825.3	-5624.5	-5624.4	-856.2	-6393.7

STREAM	92	93	94	95	96	G1	G2	G3	G5	G6	G7	G8	G9
FLOW (LB/HR)	6160	117	2595	1915	4628	4673991	4673991	4673991	4673991	4673991	4673991	4673991	4673991
TEMPERATURE (F)	217.3	305.3	432.3	629.3	213	1119.5	763	686.6	623.4	452	338.9	329.8	260.1
PRESSURE (PSIA)	16.3	72.5	352	1910.5	15	15.2	15.2	15.2	15.2	15.2	15.2	15.2	15
H (MM BTU/HR)	-35.2	-0.8	-16.8	-11.9	-29.4	-1971.2	-2426.8	-2521.6	-2599.2	-2806.6	-2940.8	-2951.6	-3033.4

A-29

Combined Cycle

IGCC Destec (E-Gas^{TM}) / HGCU / "G" Gas Turbine

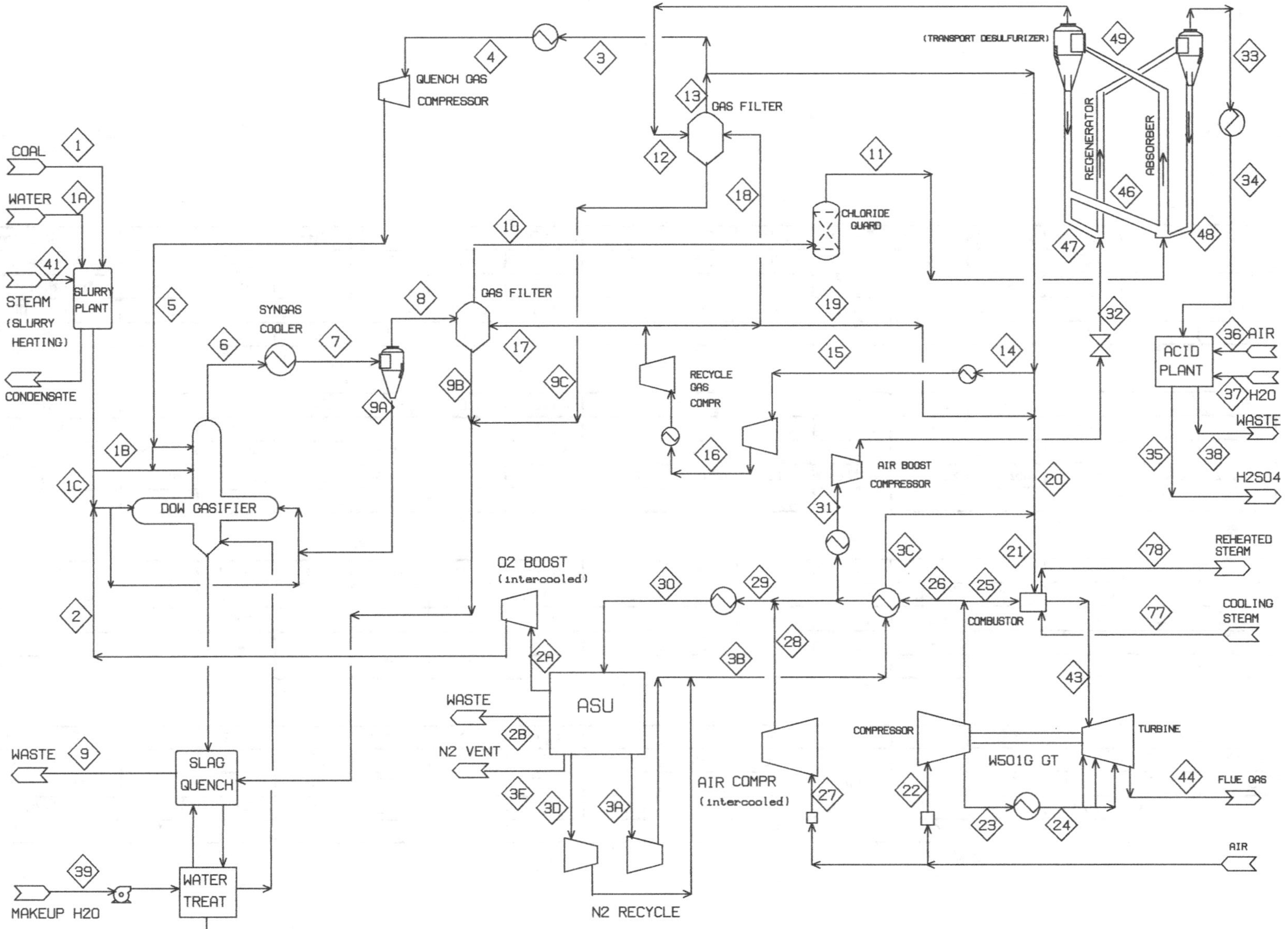

FIGURE 3A

DESTEC IGCC (HGCU/ACID PLANT/W501G GT)

A-31

DESTEC IGCC - (SYNGAS COOLER / HGCU / ACIDPLANT / 3 PRES STEAM CYCLE)

SUMMARY:

POWER	MWe
GAS TURBINE	272.6
STEAM TURBINE	171.1
MISCELLANEOUS	31
AUXILIAF	12.4
NET POWER	400.4

EFFICIENCY:	%
HHV	47.6
LHV	49.4

STREAM	1	1A	1B	1C	2A	2B	2	3A	3B	3C	3D	3E	3	4	5
FLOW (LB/HR)	245353	81753	68425	258681	189517	2823	189517	260592	299636	299636	39045	303460	166931	166931	166938
TEMPERATURE (F)	59	59	350	350	60	59	204.7	62	187.3	700	60	62	1053.2	300	360.3
PRESSURE (PSIA)	14.7	14.7	465	465	92	14.7	472	91	300	294	265	91	346	336	425
H (MM BTU/HR)	-768.2	-562.6	-146	-547.8	-0.9	-19.4	4.8	-2.8	6.1	45.3	-0.3	-3.3	-356.7	-406.9	-403.2

STREAM	6	7	8	9A	9B	9C	9	39	40	41	10	11	12	13	14
FLOW (LB/HR)	668707	668707	660421	8286	875	46	36520	29732	17272	50110	667559	666840	663762	667723	13354
TEMPERATURE (F)	1900	1004	1004	1004	997.1	1053.2	200	59	200	820.1	997.1	994.1	1057	1053.2	1053.2
PRESSURE (PSIA)	412	403.8	394.5	394.5	14.7	14.7	14.7	15	15	150	382.7	366	356	346	346
H (MM BTU/HR)	-1152.2	-1408.8	-1397.9	-10.9	-1.2	-0.1	-135.4	-204.6	-116.2	-272	-1416	-1416.1	-1417.3	-1426.8	-28.5

STREAM	15	16	17	18	19	20	21	22	23	24	25	26	27	28	29
FLOW (LB/HR)	13354	13354	8013	4006	1335	488773	788409	4320000	527109	527109	3321623	457790	397907	396217	794125
TEMPERATURE (F)	300	436.2	418.3	418.3	418.3	1051.5	952.5	59	812.7	600	812.7	812.7	59	203.7	341
PRESSURE (PSIA)	336	565.6	900	900	900	345	294	14.6	282.2	276.6	282.2	282.2	14.6	278	278
H (MM BTU/HR)	-32.6	-31.9	-19.2	-9.6	-3.2	-1044.8	-999.4	-180.3	76.8	48	483.7	66.7	-16.6	6.5	30.3

STREAM	30	31	32	33	34	35	36	37	38	43	44	46	47	48	49
FLOW (LB/HR)	794125	59882	59882	62927	62927	18585	13188	3331	60858	4110031	4637140	4390982	487887	484842	5542664
TEMPERATURE (F)	190	120	167	1420.4	850	100	59	59	100	2583	1125.6	1055	1055	1420.4	1057.9
PRESSURE (PSIA)	275	275.2	371	361	344	16	14.7	14.7	16	268.5	15.2	356	356	361	361
H (MM BTU/HR)	0.6	-1.9	-1.1	-5.7	-14.9	-23.3	-0.6	-22.9	-1.9	-554.2	-2259.1	-15077.6	-1675.3	-1672.7	-18166.3

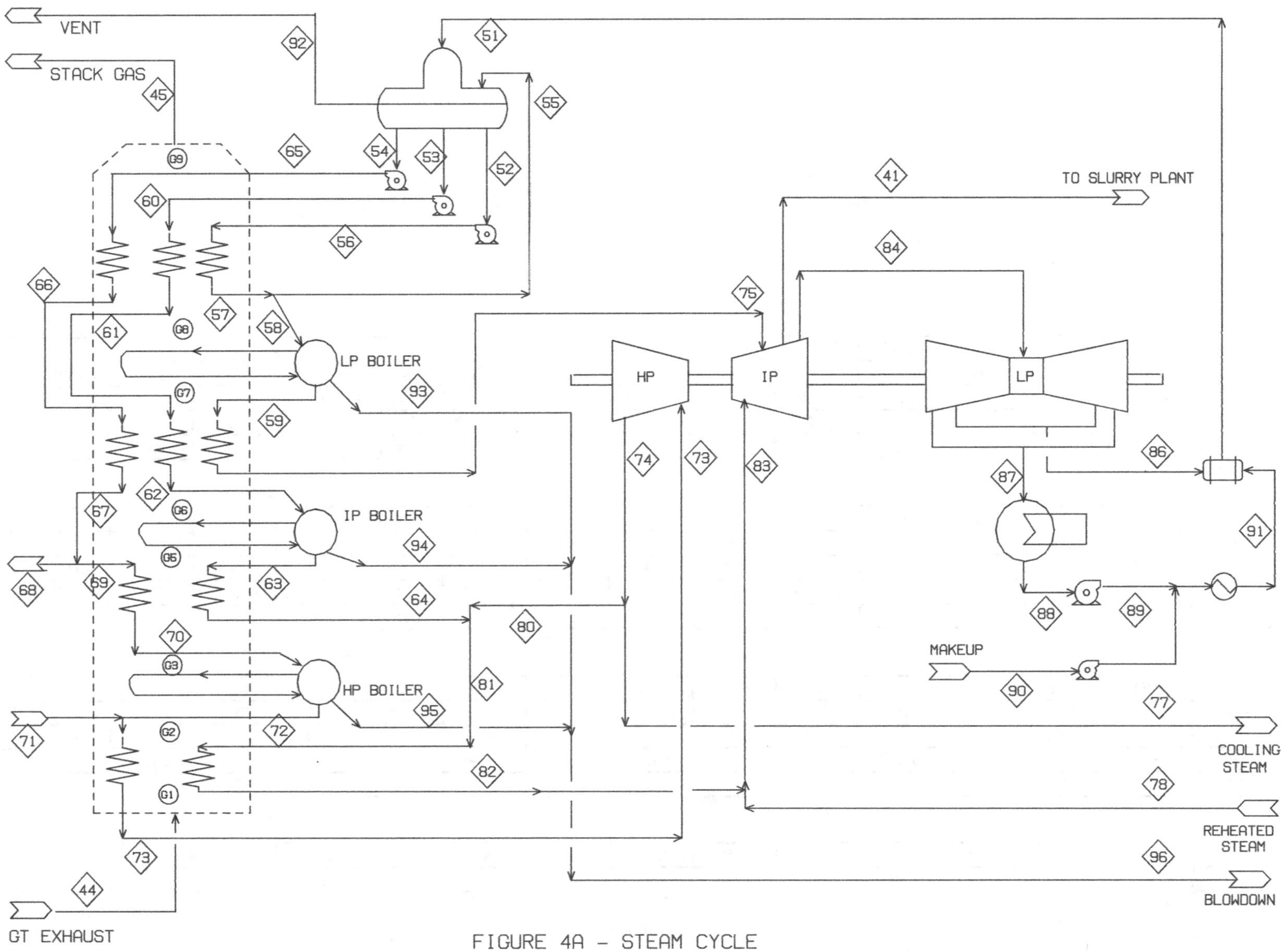

FIGURE 4A - STEAM CYCLE

DESTEC IGCC (HGCU/ACID PLANT/W501G GT)

A-33

DESTEC IGCC - (SYNGAS COOLER / HGCU / ACID PLANT / 3 PRES STEAM CYCLE)

STEAM CYCLE / HRSG PROCESS STREAMS

STREAM	41	44	45	51	52	53	54	55	56	57	58	59	60	61
FLOW (LB/HR)	50110	4637140	4637140	950123	262206	263059	669625	250771	262206	262206	11435	11321	263059	263059
TEMPERATURE (F)	820.1	1125.6	258.2	205	217.3	217.3	217.3	286	217.4	286	286	305.3	218.1	286
PRESSURE (PSIA)	150	15.2	15	17	16.3	16.3	16.3	76.3	80.3	76.3	76.3	72.5	410.6	390
H (MM BTU/HR)	-272	-2259.1	-3333.4	-6359.2	-1751.7	-1757.4	-4473.6	-1657.8	-1751.7	-1733.4	-75.6	-64.4	-1757	-1738.9

STREAM	62	63	64	65	66	67	68	69	70	71	72	73	74	75
FLOW (LB/HR)	263059	260428	260428	669625	669625	669625	437666	231959	231959	437666	229639	667305	667305	11321
TEMPERATURE (F)	420	432.3	620	221.2	286	420	420	420	620	635	629.3	1049.3	606.2	420
PRESSURE (PSIA)	370.5	352	350	2345.6	2228.3	2116.9	2116.9	2116.9	2011.1	1911	1910.5	1800	350	69.5
H (MM BTU/HR)	-1701.8	-1474.7	-1443.8	-4467.5	-4424.1	-4330.9	-2830.7	-1500.2	-1443.1	-2498.9	-1313.9	-3573.8	-3704.6	-63.7

STREAM	77	78	80	81	82	83	84	86	87	88	89	90	91	92
FLOW (LB/HR)	70000	70000	597305	857733	857733	927733	888944	51176	837768	837768	837768	61179	898947	6004
TEMPERATURE (F)	606.2	1055.4	606.2	610.4	1050	1050.4	481.9	350	88.8	87.9	87.9	80	145.7	217.3
PRESSURE (PSIA)	350	342	350	350	342	342	35	17	0.7	0.7	40	14.7	17	16.3
H (MM BTU/HR)	-388.6	-371.8	-3316	-4759.8	-4558.6	-4930.5	-4969	-289.2	-4894.6	-5705.3	-5705.2	-417.1	-6070.1	-34.3

STREAM	93	94	95	96	G1	G2	G3	G5	G6	G7	G8	G9
FLOW (LB/HR)	114	2631	2320	5065	4637140	4637140	4637140	4637140	4637140	4637140	4637140	4637140
TEMPERATURE (F)	305.3	432.3	629.3	213	1125.6	782.5	690.3	618.8	445.1	335	326.1	258.2
PRESSURE (PSIA)	72.5	352	1910.5	15	15.2	15.2	15.2	15.2	15.2	15.2	15.2	15
H (MM BTU/HR)	-0.8	-17	-14.4	-32.1	-2259.1	-2699.3	-2814.1	-2902.1	-3112.3	-3243.2	-3253.7	-3333.4

Combined Cycle

IGCC Destec (E-Gas™) / CGCU / "G" Gas Turbine / CO_2 Capture

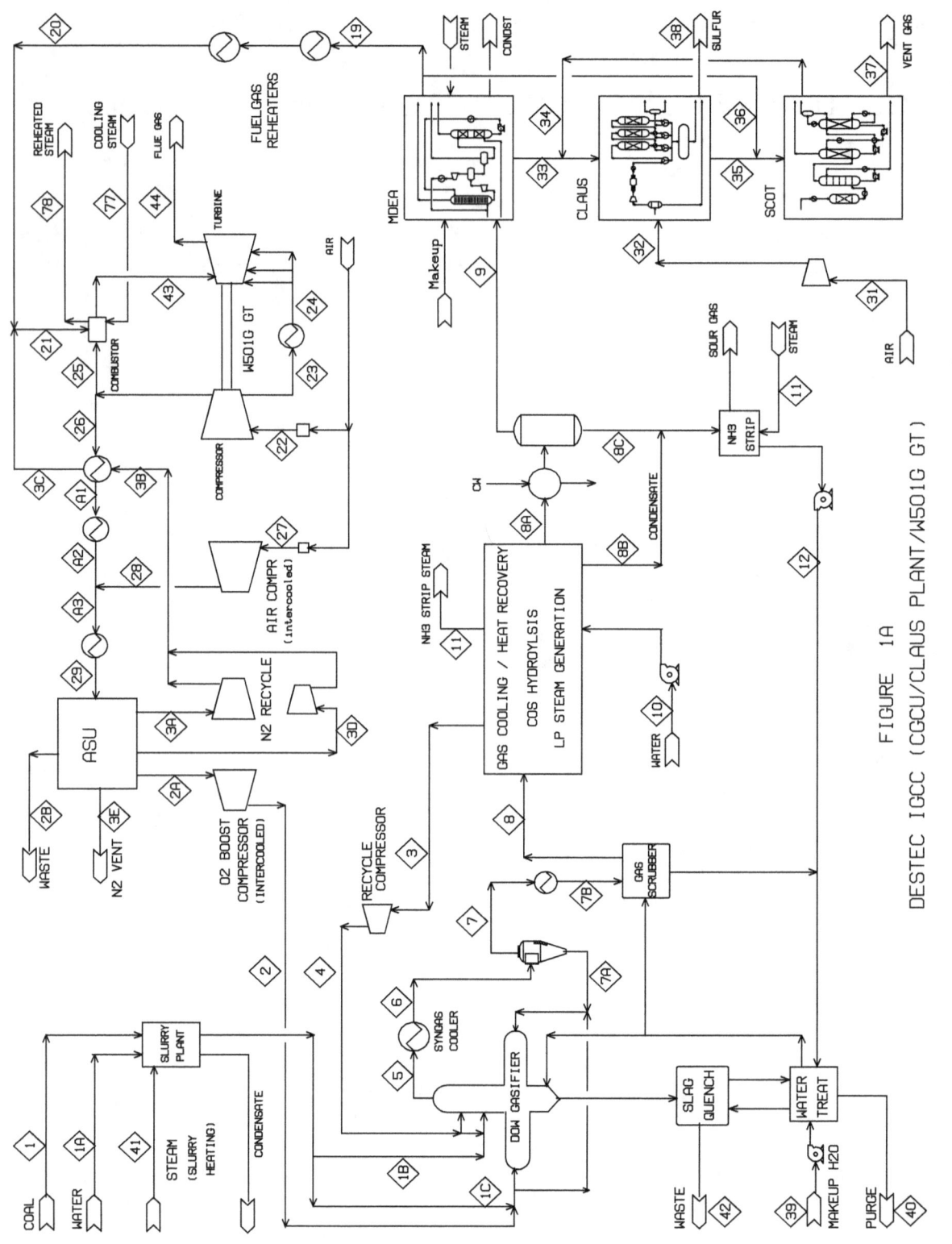

FIGURE 1A

DESTEC IGCC (CGCU/CLAUS PLANT/W501G GT)

DESTEC IGCC - (SYNGAS COOLER / CGCU / CLAUS PLANT / 3 PRES STEAM CYCLE)

SUMMARY:

POWER	MWe
GAS TURBINE	272.8
STEAM TURBINE	172.2
MISCELLANEOUS	32
AUXILIAF	12.4
NET POWER	400.6

EFFICIENCY:	%
HHV	45
LHV	46.7

STREAM	1	1A	1B	1C	2A	2B	2	3A	3B	3C	3D	3E	3	4
FLOW (LB/HR)	260226	86709	72573	274362	197846	2990	197846	317975	358735	358735	40761	270868	141102	141103
TEMPERATURE (F)	59	59	350	350	60	59	204.7	62	189.3	700	60	62	304.6	333.8
PRESSURE (PSIA)	14.7	14.7	465	465	92	14.7	472	91	300	294	265	91	378	425
H (MM BTU/HR)	-814.7	-596.7	-153.1	-574.8	-0.9	-20.3	5	-3.5	7.4	54.1	-0.3	-3	-370.7	-369.2

STREAM	5	6	7	7A	7B	8	8A	8B	8C	9	10	11	12	19
FLOW (LB/HR)	670129	670129	661340	8788	661340	705510	472085	92323	11380	460705	45000	45000	102871	424837
TEMPERATURE (F)	1900	650	649.9	649.9	415	304.2	190	232.2	101.9	103	59	280	213	116
PRESSURE (PSIA)	412	403.8	394.5	394.5	390	380	354	354	20	349	14.7	37	470	340
H (MM BTU/HR)	-1164.4	-1519.1	-1506.8	-12.4	-1568	-1853.6	-978	-617.7	-73.3	-931.2	-309.7	-255.7	-690.7	-847.5

STREAM	20	21	22	23	24	25	26	A1	A2	A3	27	28	29	31
FLOW (LB/HR)	424837	783573	4320000	527109	527109	3363310	416102	416102	416102	830440	416102	414338	830440	14107
TEMPERATURE (F)	589.7	629.2	59	813.3	600	813.3	813.3	370.4	216	210	59	203.9	190	59
PRESSURE (PSIA)	330	294	14.6	282.2	276.6	282.2	282.2	280.2	278	278	14.6	278	277	14.7
H (MM BTU/HR)	-770.3	-716.2	-180.3	76.7	47.9	489.6	60.6	13.8	-2.1	4.6	-17.4	6.7	0.5	-0.6

STREAM	32	33	34	35	36	37	38	39	40	41	42	43	44	45
FLOW (LB/HR)	14107	32346	1976	42121	6755	46900	6307	38680	87473	53025	34450	4146881	4673991	4673991
TEMPERATURE (F)	161.2	142.1	70	424	116	70	285	59	200	820.1	200	2582.2	1119.5	261
PRESSURE (PSIA)	25	18.5	17.5	26.7	340	17.5	14.7	14.7	15	150	15	268.5	15.2	14.7
H (MM BTU/HR)	-0.2	-86.3	-6.9	-109.4	-13.5	-126.8	-0.7	-266.2	-585.9	-287.8	-114.7	-265.9	-1971.2	-3032.3

STEAM CYCLE

DESTEC IGCC (CGCU/CLAUS PLANT/W501G GT)

A-38

DESTEC IGCC - (SYNGAS COOLER / CGCU / CLAUS PLANT / 3 PRES STEAM CYCLE)

STEAM CYCLE / HRSG PROCESS STREAMS.

STREAM	41	44	45	51	52	53	54	55	56	57	58	59	60	61
FLOW (LB/HR)	5302.5	4673991	4673991	974779	269014	259543	697343	257282	269014	269014	11732	11615	259543	259543
TEMPERATURE (F)	820.1	1119.5	261	203.8	217.3	217.3	217.3	286	217.4	286	286	305.3	218.1	286
PRESSURE (PSIA)	150	15.2	14.7	17	16.3	16.3	16.3	76.3	80.3	76.3	76.3	72.5	410.6	390
H (MM BTU/HR)	-287.8	-1971.2	-3032.3	-6525.4	-1797.2	-1733.9	-4658.8	-1700.9	-1797.1	-1778.4	-77.6	-66	-1733.5	-1715.7

STREAM	62	63	64	65	66	67	68	69	70	71	72	73	74	75
FLOW (LB/HR)	259543	256948	256948	697343	697343	697343	505841	191503	191503	505841	189588	695428	695428	11615
TEMPERATURE (F)	420	432.3	620	221.2	286	420	420	620	620	635	629.3	1050	606.7	420
PRESSURE (PSIA)	370.5	352	350	2345.6	2228.3	2116.9	2116.9	2011.1	2011.1	1910.5	1910.5	1800	350	69.5
H (MM BTU/HR)	-1679.1	-1455	-1424.5	-4652.4	-4607.2	-4510.2	-3271.6	-1191.4	-1191.4	-2888.1	-1084.7	-3724.1	-3860.5	-65.3

STREAM	77	78	80	81	82	83	84	85	86	87	88	89	90	91
FLOW (LB/HR)	70000	70000	625428	882376	882376	952376	849203	23299	61763	825904	825904	825904	125576	951480
TEMPERATURE (F)	606.7	1055.9	606.7	610.6	1050	1050.4	482	350	596.5	88.8	87.9	87.9	80	178.3
PRESSURE (PSIA)	350	342	350	350	342	342	35	17	60	0.7	0.7	40	14.7	17
H (MM BTU/HR)	-388.6	-371.8	-3472	-4896.4	-4689.6	-5061.4	-4746.9	-131.7	-341.9	-4825.3	-5624.5	-5624.4	-856.2	-6393.7

STREAM	92	93	94	95	96	G1	G2	G3	G5	G6	G7	G8	G9
FLOW (LB/HR)	6160	117	2595	1915	4628	4673991	4673991	4673991	4673991	4673991	4673991	4673991	4673991
TEMPERATURE (F)	217.3	305.3	432.3	629.3	213	1119.5	763	686.6	623.4	452	338.9	329.8	260.1
PRESSURE (PSIA)	16.3	72.5	352	1910.5	15	15.2	15.2	15.2	15.2	15.2	15.2	15.2	15
H (MM BTU/HR)	-35.2	-0.8	-16.8	-11.9	-29.4	-1971.2	-2426.8	-2521.6	-2599.2	-2806.6	-2940.8	-2951.6	-3033.4

A-39

Combined Cycle

IGCC Shell / CGCU / "G" Gas Turbine

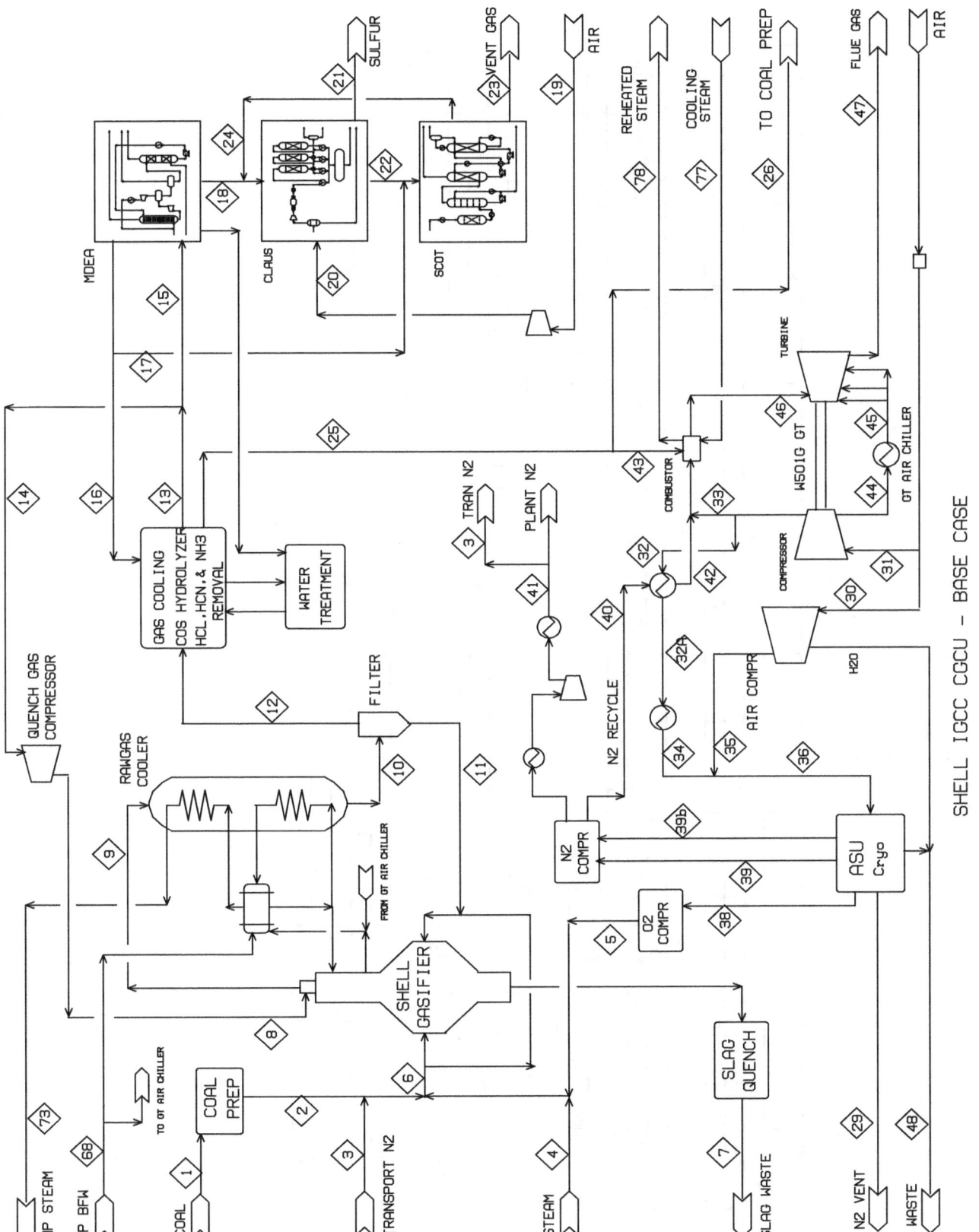

SHELL IGCC CGCU - BASE CASE

A-41

FIGURE 1B

SHELL IGCC CGCU - BASE CASE

SUMMARY :

POWER	MWe
GAS TURBINE	272.3
STEAM TURBINE	188.9
MISCELLANEOUS	35.5
AUXILIARY (3%)	12.8
PLANT TOTAL	412.8

EFFICIENCY:	%
HHV	45.7
LHV	47.4

STREAM	1	2	3	4	5	6	7	8	9	10	11	12	13	14	15
FLOW (LB/HR)	264263	248089	18971	7214	213207	488857	26747	194116	656226	656226	1408	654818	647053	194116	452937
TEMPERATURE (F)	59	59	104	694	204.7	144.9	300	123.9	1843.7	640	640	640	100	327.5	100
PRESSURE (PSIA)	14.7	14.7	400	500	472	370	14.7	370	352.5	347.5	347.5	342.5	327.5	327.5	327.5
H (MM BTU/HR)	-972.6	-155.9	0.1	-39.8	5.4	-193	-62.3	-311.6	-669.3	-964.7	-3.1	-961.6	-1043.9	-313.2	-730.7

STREAM	16	17	18	19	20	21	22	23	24	25	26	29	30	31	32
FLOW (LB/HR)	435249	7243	12078	14529	14529	6496	20730	27354	619	435249	3174	234788	448410	320000	448410
TEMPERATURE (F)	116	116	160.3	59	161.2	285	430.8	70	70	600	600	62	59	59	813.3
PRESSURE (PSIA)	323	323	18.5	14.7	25	14.7	26.7	17.5	17.5	318	318	91	14.6	14.6	282.2
H (MM BTU/HR)	-701.6	-11.7	-24.2	-0.6	-0.2	-0.7	-44.2	-62.4	-1.7	-628.4	-4.6	-2.6	-18.7	-180.3	65.3

STREAM	32A	33	34	35	36	38	39	39B	40	41	42	43	44	45	46
FLOW (LB/HR)	448410	3331003	448410	446508	894918	213207	399775	43925	415244	28456	415244	432075	527109	527109	4178319
TEMPERATURE (F)	334.1	813.3	190	203.9	196.9	60	62	60	198.7	105	712	600	813.3	600	2583.1
PRESSURE (PSIA)	280.2	282.2	278	278	278	92	91	265	300	401.8	294	318	282.2	276.6	268.5
H (MM BTU/HR)	10.8	484.8	-5.2	7.3	2.1	-1	-4.4	-0.3	9.3	0.1	63.8	-623.8	76.7	47.9	-114.6

STREAM	47	48	68	73	77	78
FLOW (LB/HR)	4705428	5124	440022	440022	70000	70000
TEMPERATURE (F)	1117.5	59	420	1050	606.2	1055.4
PRESSURE (PSIA)	15.2	15	2116.9	1815	350	342
H (MM BTU/HR)	-1818.1	-35	-2845.9	-2356.5	-388.6	-371.8

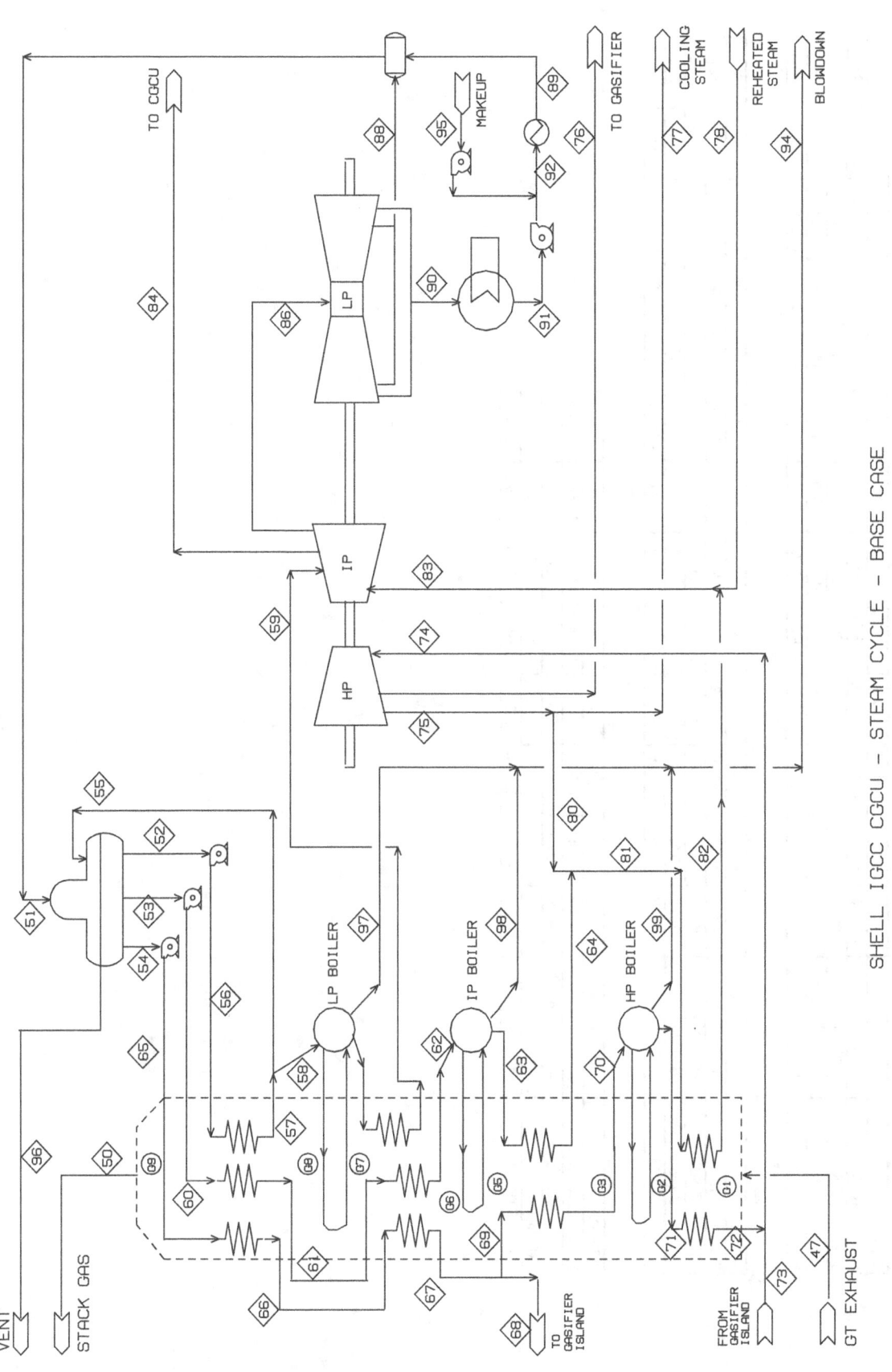

SHELL IGCC CGCU – STEAM CYCLE – BASE CASE

A-43

Shell IGCC CGCU - Steam Cycle /HRSG Streams

STREAM	47	50	51	52	53	54	55	56	57	58	59	60	61	62	63
FLOW (LB/HR)	4705428	4705428	1034798	285578	199288	816516	273123	285578	285578	12454	12330	199288	199288	199288	197295
TEMPERATURE (F)	1117.5	260	205	217.3	217.3	217.3	286	217.4	286	286	420	218.1	286	420	432.3
PRESSURE (PSIA)	15.2	14.7	17	16.3	16.3	16.3	76.3	80.3	76.3	76.3	70.5	410.6	390	370.5	352
H (MM BTU/HR)	-1818.1	-2876	-6925.9	-1907.9	-1331.4	-5454.9	-1805.6	-1907.8	-1887.9	-82.3	-69.3	-1331	-1317.4	-1289.3	-1117.2

STREAM	64	65	66	67	68	69	70	71	72	73	74	75	76	77	78
FLOW (LB/HR)	197295	816516	816516	816516	440022	376494	376494	372729	372729	440022	812751	805536	7214	70000	70000
TEMPERATURE (F)	620	221.1	286	420	420	420	620	629.3	1050	1050	1049.3	606.2	695.7	606.2	1055.4
PRESSURE (PSIA)	350	2345.6	2228.3	2116.9	2116.9	2116.9	2011.1	1910.5	1815	1815	1800	350	510	350	342
H (MM BTU/HR)	-1093.8	-5447.6	-5394.5	-5281	-2845.9	-2435	-2342.3	-2132.6	-1996.2	-2356.5	-4352.7	-4472	-39.8	-388.6	-371.8

STREAM	80	81	82	83	84	86	88	89	90	91	92	94	95	96	97
FLOW (LB/HR)	735536	932832	932832	1002832	86350	928812	50648	984150	878164	878164	984150	5882	105986	6540	125
TEMPERATURE (F)	606.2	609.1	1050	1050.4	600	485.1	352.8	151.6	88.8	87.9	87	213	80	217.3	305.3
PRESSURE (PSIA)	350	350	342	342	60	35	17	17	0.7	0.7	17	15	14.7	16.3	72.5
H (MM BTU/HR)	-4083.4	-5177.2	-4957.8	-5329.6	-477.9	-5190.4	-286.1	-6639.6	-5129.8	-5980.4	-6702.9	-37	-722.6	-37.4	-0.8

STREAM	98	99	G1	G2	G3	G5	G6	G7	G8	G9
FLOW (LB/HR)	1993	3765	4705428	4705428	4705428	4705428	4705428	4705428	4705428	4705428
TEMPERATURE (F)	432.3	629.3	1117.5	839.9	690.3	595.5	463.5	343.6	333.9	259.9
PRESSURE (PSIA)	352	1910.5	15.2	15.2	15.2	15.2	15.2	15.2	15.2	15
H (MM BTU/HR)	-12.9	-23.4	-1818.1	-2174	-2360.4	-2476.5	-2635.7	-2778.1	-2789.5	-2876.1

A-44

Combined Cycle

IGCC Shell / CGCU / "G" Gas Turbine / CO_2 Capture

Shell-based Gasification Combined Cycle Plant with Hydrogen and Carbon Dioxide

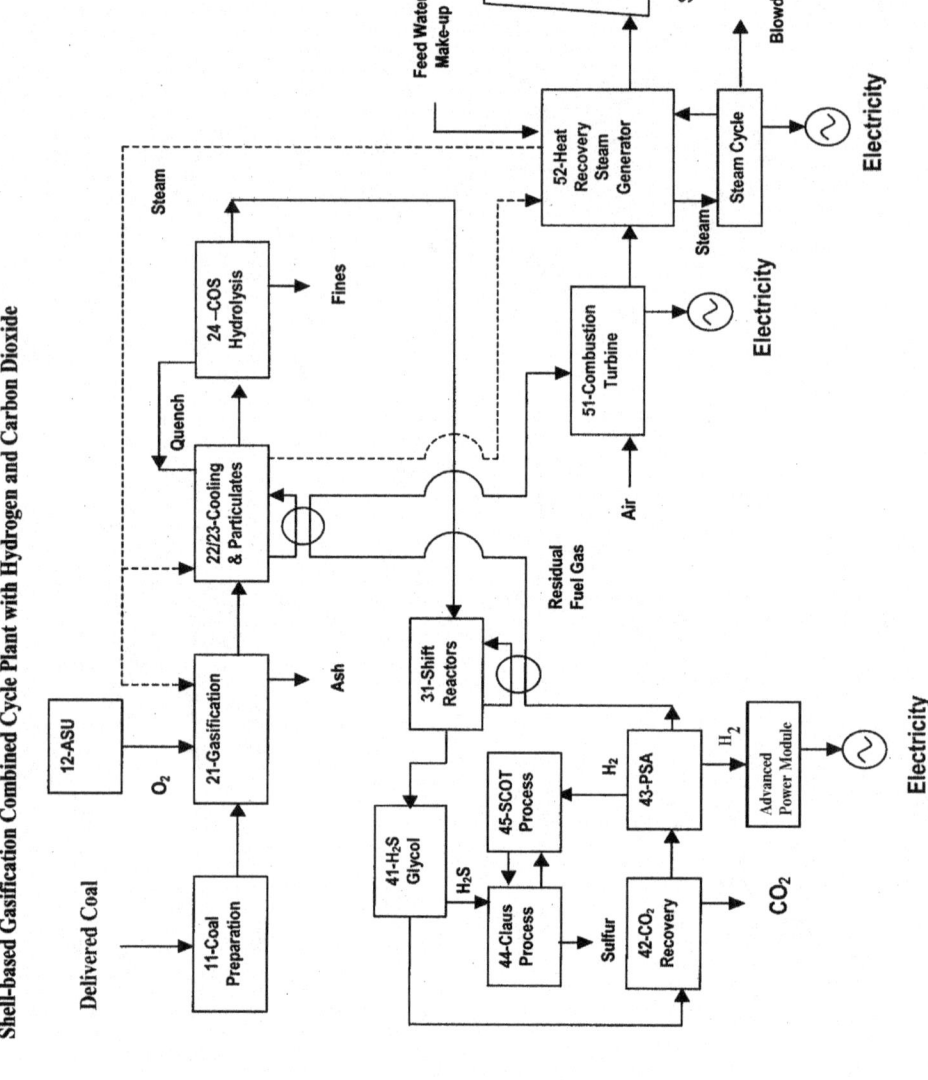

Material & Energy Balance

Results by ANL : (J. Molburg, R. Doctor , N. Brockmeier)

Are storied on

NETL/Gasification Technologies team website (Publications) :
http://www.netl.doe.gov/coalpower/gasification/pubs/pdf/igcc-co2.pdf

Hydraulic Air Compression (HAC)

Natural Gas HAC - No CO_2 Capture

HYDRAULIC AIR COMPRESSION CYCLE - NATURAL GAS - NO CO2 SEQUESTRATION

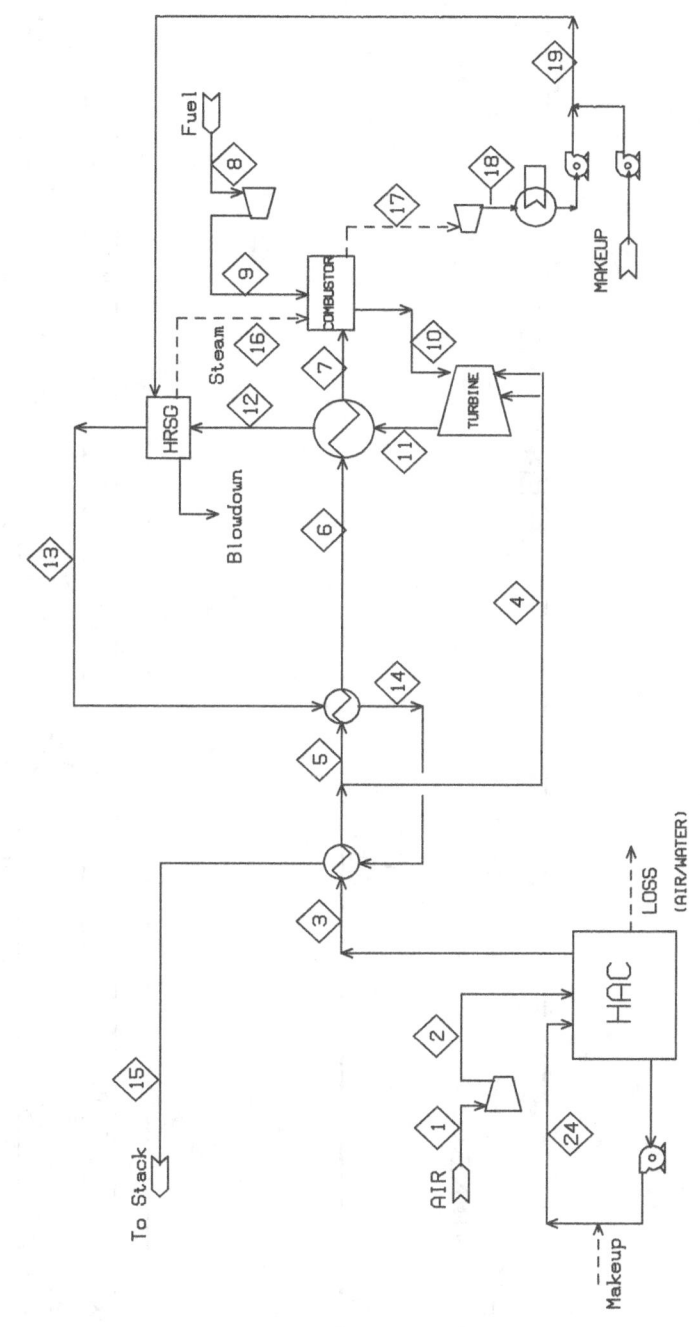

HYDRAULIC AIR COMPRESSION CYCLE - NATURAL GAS - NO CO2 SEQUESTRATION

	MWe
GT EXPANDER	323.5
STEAM TURBINE	6.1
HAC	170.7
MISC/AUX	6.6
NET POWER	323.5

EFFICIENCY		%
	LHV	53.2
	HHV	48.1

STREAM ID	1	2	3	4	5	6	7	8	9	10
Mass Flow lb/hr	4203605	4203605	4196597	263544	3933053	3933053	3933053	96465	96465	4029509
Temperature F	59	66	60	100	100	175	950	60	191.6	2583
Pressure psi	14.7	15.3	282	277.9	277.9	273.8	273.8	150	350	268.5
H MMBtu/hr	-175.6	-168.5	-209.9	-10.1	-150.4	-59.9	713.2	-194	-187.4	498.5

STREAM ID	11	12	13	14	15	16	17	18	19	24
Mass Flow lb/hr	4293063	4293063	4293063	4293063	4293063	80000	80000	80000	80808	4690215160
Temperature F	1127.9	479.4	400	318	273	265	699.1	131.2	96.1	59
Pressure psi	15.2	14.9	14.9	14.8	14.7	35	30	1	40	58.6
H MMBtu/hr	-1225.5	-1983.2	-2071.8	-2162.3	-2211.7	-455.7	-438.7	-459.8	-549.6	-3.21E+07

Hydraulic Air Compression (HAC)

Natural Gas HAC - CO_2 Capture

HYDRAULIC AIR COMPRESSION CYCLE - NATURAL GAS - CO2 SEQUESTRATION

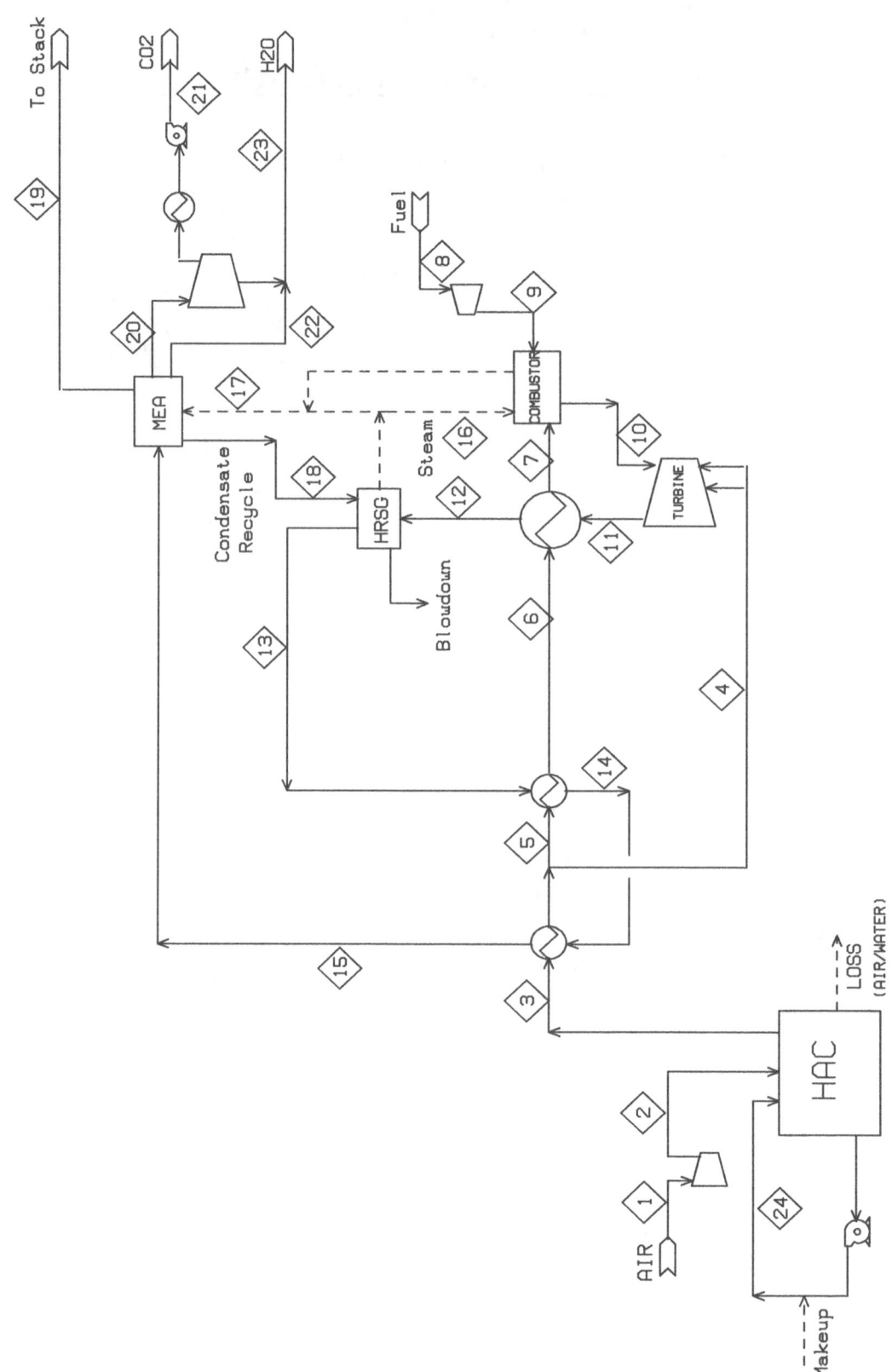

HYDRAULIC AIR COMPRESSION CYCLE - NATURAL GAS - CO2 SEQUESTRATION

	MWe		EFFICIENCY	%
GT EXPANDER	498.8		LHV	43.8
HAC	170.7		HHV	39.6
CO2 RECOVERY	11.4			
MISC/AUX	16.5			
NET POWER	300.2			

STREAM ID	1	2	3	4	5	6	7	8	9	10
Mass Flow lb/hr	4203605	4203605	4196587	263544	3933043	3933043	3933043	108611	108611	4041657
Temperature F	59	66	60	100	100	275	725	60	191.6	2583
Pressure psi	14.7	15.3	282	277.9	277.9	273.8	277	150	350	268.5
H MMBtu/hr	-175.6	-168.5	-209.9	-10.1	-150.4	37.6	482.7	-218.5	-210.9	243.1

STREAM ID	13	14	15	16	17	18	19	20	21	22
Mass Flow lb/hr	4305211	4305211	4305211	80000	471902	471902	3901949	277066	270109	4305211
Temperature F	332.4	161.9	119.1	428	428	250.3	100	140	103.6	100
Pressure psi	14.9	14.8	14.7	35	35	45	14.7	25.7	3000	14.7
H MMBtu/hr	-2429.8	-2617.8	-2667.2	-449.3	-2650.1	-3136.9	-880.6	-1075.4	-1065.9	-2821.7

Hydraulic Air Compression (HAC)

- Destec (E-Gas^{TM}) / CGCU / "G" GT / No CO_2 Capture

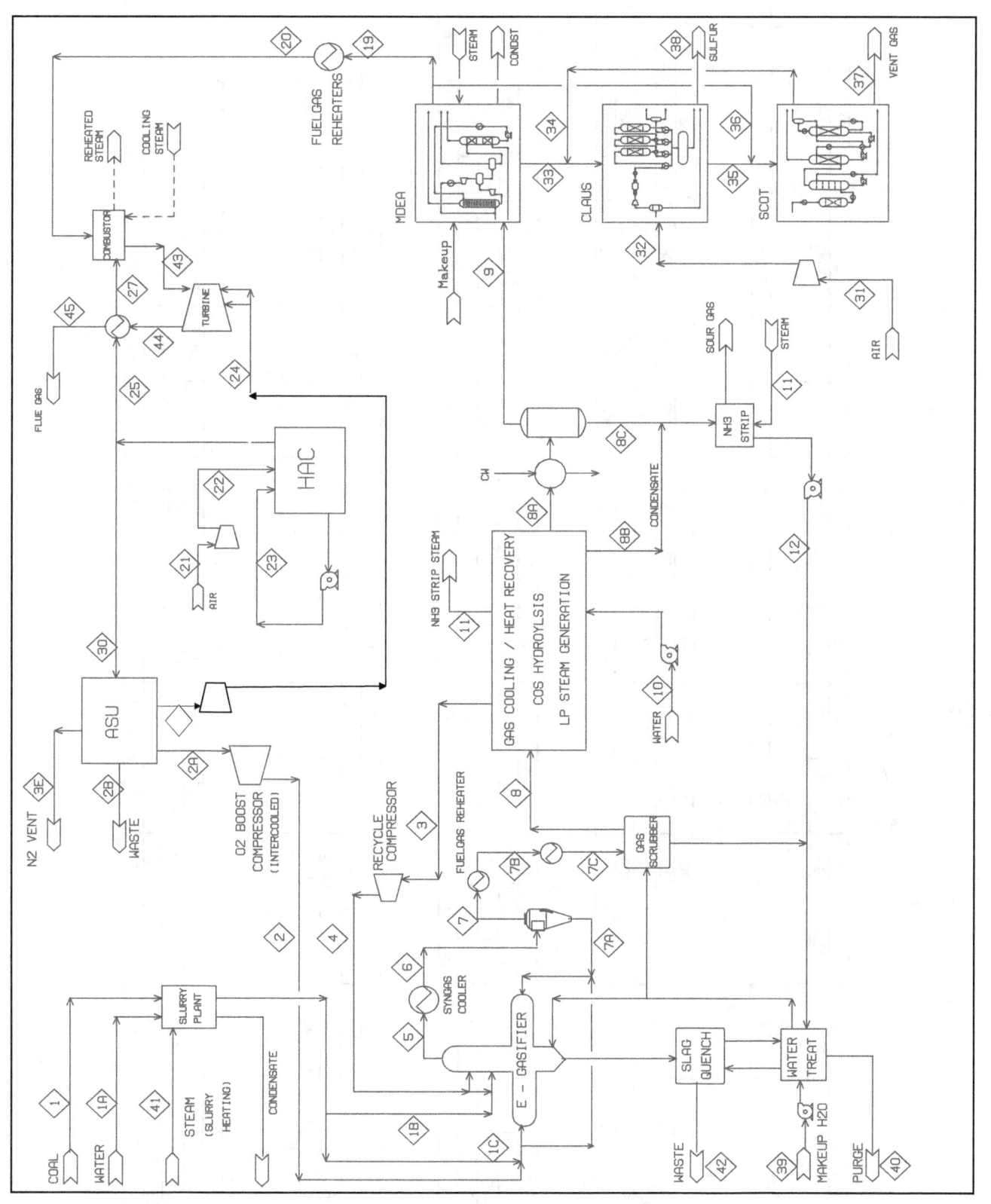

A-55

CASE 3

SUMMARY - COAL POWERED HAC PROCESS (NO CO2 CAPTURE)

	MWe
GT EXPANDER	499.1
STEAM TURBINE	30.9
HAC	184.1
MISC / AUX	20.0
NET POWER	325.9

EFFICIENCY		%
	LHV	43.8
	HHV	42.3

STREAM ID	1	1A	1B	1C	2A	2B	2	3E	3A	3	4	5	6	7
FLOW (lb/hr)	225480	75132	62883	237728	171462	865	171462	282097	263544	122547	122543	580965	580965	573350
TEMPERATURE °F	59	59	350	350	60	59	204.7	61.5	61.5	305.3	334.5	1900	1100	1100
PRESSURE psi	14.7	14.7	465	465	92	14.7	472	91	91	378	425	412	403.8	394.5
H (MMBtu/hr)	-705.9	-517	-132.6	-498.1	-0.8	-6	4.3	-3	-2.8	-322.9	-321.5	-1010.6	-1212	-1202.2

STREAM ID	7A	7B	7C	8	8A	8B	8C	9	10	11	12	19	20	21
FLOW (lb/hr)	7615	573350	573350	612734	409017	81170	9858	399159	45000	45000	90308	367192	367192	4537440
TEMPERATURE °F	1100	820	415	304.9	190	232.4	101.9	103	59	280	213.4	116	584.9	59
PRESSURE psi	394.5	390	390	380	354	354	20	349	14.7	37	470	340	330	14.7
H (MMBtu/hr)	-9.8	-1268.2	-1360.6	-1614.4	-847.5	-543.1	-63.5	-806.9	-309.7	-255.7	-606.3	-732.6	-666.6	-189.4

STREAM ID	22	23	24	25	27	30	31	32	33	34	35	36	37	38
FLOW (lb/hr)	4537440	5.06E+09	263544	3816000	3816000	718166	12185	12185	28014	1732	36482	6755	41505	5448
TEMPERATURE °F	65.8	59	100	60	1090	60	59	161.2	142.1	70	424	116	70	285
PRESSURE psi	15.3	58.6	105.7	282	14.7	275	14.7	25	18.5	17.5	26.7	340	17.5	14.7
H (MMBtu/hr)	-182	-3.46E+07	-0.3	-190.7	833	-15	-0.5	-0.2	-74.8	-6	-94.9	-13.5	-111.7	-0.6

STREAM ID	39	40	41	42	43	44	45	51	52	53	54	55	73
FLOW (lb/hr)	33516	75855	45945	29850	4183478	4447021	4447021	197294	199591	199591	199591	197595	197595
TEMPERATURE °F	59	200	821.6	200	2581.4	1141.2	268.5	205	217.3	222.5	620	629.3	1050
PRESSURE psi	14.7	15	150	15	268.5	15.2	14.7	17	16.3	2345.6	2011.1	1910.5	1800
H (MMBtu/hr)	-230.7	-507.7	-249.4	-99.4	141.1	-1586.3	-2611	-1320.5	-1333.4	-1331.3	-1241.7	-1130.6	-1058.1

STREAM ID	74	77	78	80	82	83	84	85	86	87	88	89	90
FLOW (lb/hr)	197595	70000	70000	127595	127595	197595	98138	3300	53512	94838	94838	94838	102456
TEMPERATURE °F	606.7	606.7	1055.9	606.7	1050	1052.1	485.2	352.9	600.1	88.8	87.9	87.9	80
PRESSURE psi	350	350	342	350	342	342	35	17	60	0.7	0.7	40	14.7
H (MMBtu/hr)	-1096.9	-388.6	-371.8	-708.3	-678.1	-1049.9	-548.4	-18.6	-296.1	-554	-645.9	-645.8	-698.5

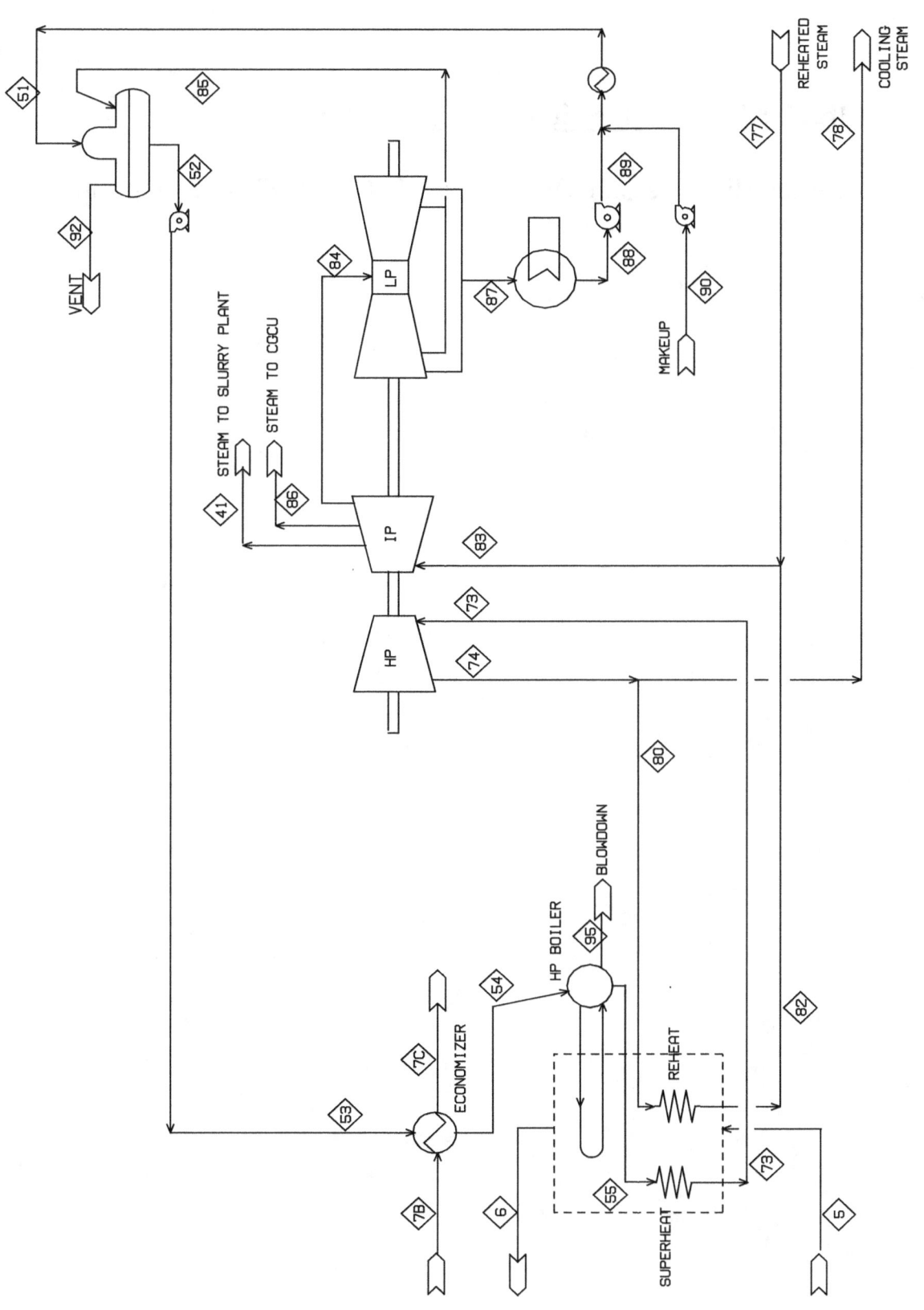

VENT

51

85

52

92

STEAM TO SLURRY PLANT

STEAM TO CGCU

41

86

84

LP

IP

83

87

88

89

90

MAKEUP

77

REHEATED STEAM

78

COOLING STEAM

73

HP

74

80

BLOWDOWN

95

HP BOILER

54

ECONOMIZER

7C

53

82

7B

REHEAT

6

SUPERHEAT

55

73

5

A-57

Hydraulic Air Compression (HAC)

- Destec High Pressure (E-Gas^{TM}) / HGCU / "G" GT / CO_2 Capture

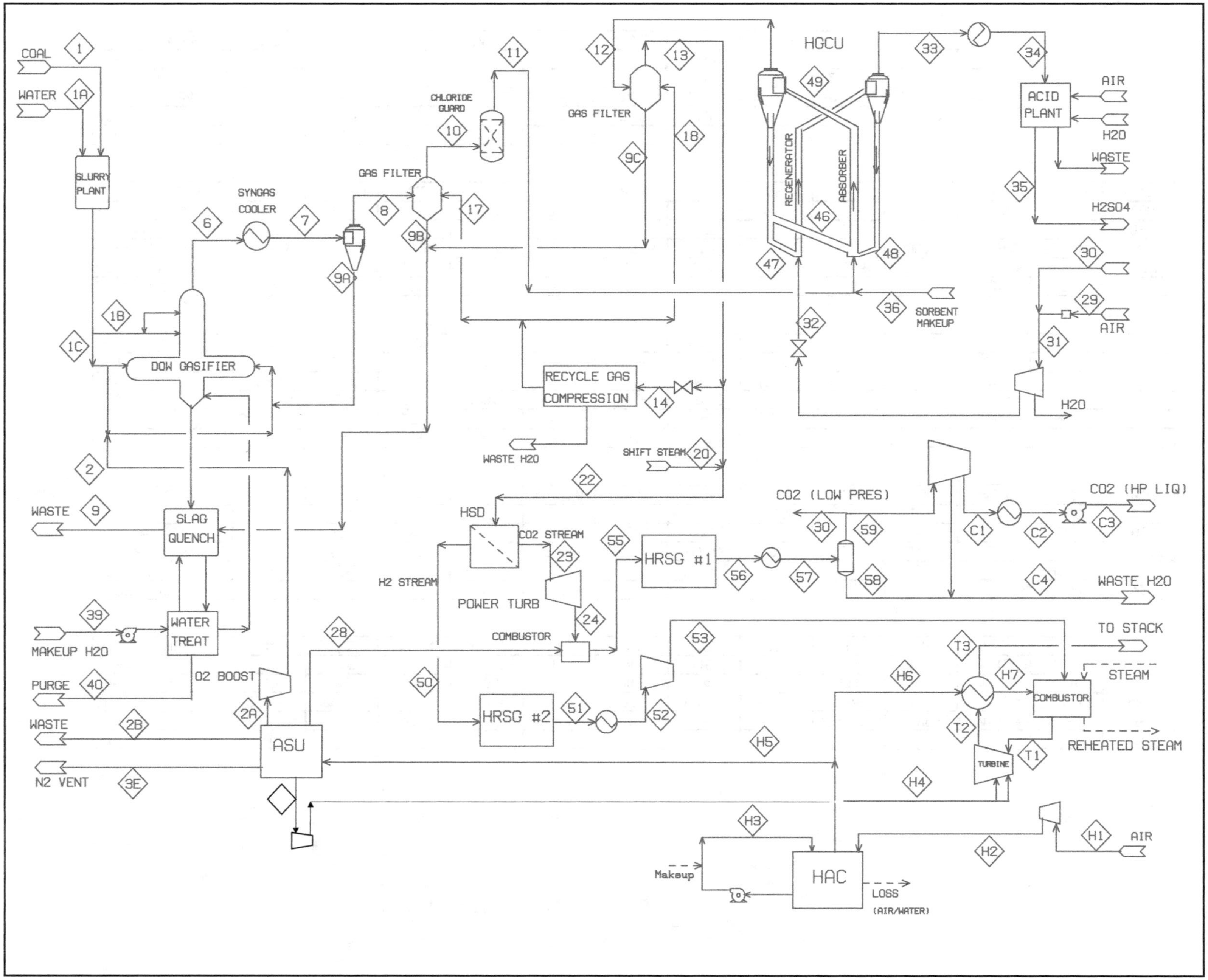

COAL
1
WATER
1A
SLURRY PLANT
1B
1C
DOW GASIFIER
2
6
SYNGAS COOLER
7
9A
8
GAS FILTER
9B
17
10
CHLORIDE GUARD
11
12
13
GAS FILTER
9C
18
HGCU
49
REGENERATOR
ABSORBER
46
47
48
33
34
ACID PLANT
AIR
H2O
WASTE
35
H2SO4
30
29
AIR
31
H2O
32
36
SORBENT MAKEUP
RECYCLE GAS COMPRESSION
14
WASTE H2O
SHIFT STEAM
20
22
HSD
CO2 STREAM
23
H2 STREAM
POWER TURB
24
COMBUSTOR
55
HRSG #1
56
57
58
30
59
CO2 (LOW PRES)
C1
C2
C3
CO2 (HP LIQ)
C4
WASTE H2O
WASTE
9
SLAG QUENCH
39
MAKEUP H2O
WATER TREAT
PURGE
40
O2 BOOST
2A
28
WASTE
2B
ASU
N2 VENT
3E
50
HRSG #2
51
52
53
TO STACK
STEAM
T3
H6
H7
COMBUSTOR
T2
REHEATED STEAM
TURBINE
T1
H5
H4
H3
Makeup
HAC
LOSS
(AIR/WATER)
H2
H1
AIR

CASE 4
HYDRAULIC AIR COMPRESSION CYCLE - COAL SYNGAS - CO2 SEQUESTRATION

	MWe
GT EXPANDER	501.7
CO2 EXPANDER	58.5
STEAM TURBINE	47.6
HAC	204.1
CO2 SEQ	28.2
H2 COMPR	26.1
MISC / AUX	36.9
NET POWER	312.4

EFFICIENCY	%
LHV	35.2
HHV	33.9

STREAM ID	1	1A	1B	1C	2A	2B	2	3E	6	7	8	9A	9B
ASPEN ID	COLIN	WAT1	COLB	COLA	GO2A	7	GOXYG	9	DRXROUT	RAWPRD	DRAWGAZ	FNES	16
Mass Flow lb/hr	269657	89852	75203	284306	179573	224	179573	574110	522761	522761	513654	9107	961
Temperature F	59	59	350	350	60	80.1	294.5	61.5	1904.8	1110	1110	1110	1098.2
Pressure psi	14.7	14.7	1078	1078	92	14.6	1150	91	1034	1024	1019	1019	14.7
H MMBtu/hr	-844.3	-618.7	-167.4	-626.7	-0.8	-1.5	7.8	-33.1	-831.9	-1016.9	-1005.2	-11.7	-1.2

STREAM ID	9C	9	39	40	10	11	12	13	14	17	18	20	22
ASPEN ID	19	WSTSOL	MWATG	PURGE	17	18	20	26	21	25	24C	SHFSTM	TOSHF1
Mass Flow lb/hr	501	40588	32677	18983	520828	520037	517105	520991	12712	8135	4387	272791	781077
Temperature F	1129.3	200	59	200	1098.2	1094.3	1135.9	1129.3	1129.3	334.3	334.3	875	1013.5
Pressure psi	14.7	14.7	15	15	1000	985	975	965	965	1291.2	1291.2	1000	964
H MMBtu/hr	-1.9	-150.8	-225	-127.8	-1022.6	-1022.7	-1035.5	-1043.7	-25.5	-18.6	-10.1	-1482.8	-2501

STREAM ID	23	24	28	29	30	31	32	33	34	35	46	47	48
ASPEN ID	CO2RICH	S5	O2CAT	RAIR	30X	N830	39	5A	5	SACID	46	47	48
Mass Flow lb/hr	740345	740345	57449	42310	45526	87836	87145	90525	90525	20786	6440444	715605	712225
Temperature F	1391.2	555.3	60	59	60	56.7	260.9	1383.4	850	100	1134.6	1134.6	1383.4
Pressure psi	950	20.5	92	14.6	14.8	14.6	971	955	940	16	975	975	955
H MMBtu/hr	-2678.6	-2881.1	-0.3	-1.8	-174	-175.8	-168.2	-168.6	-181.7	-25.9	-22020.5	-2446.7	-2448.3

STREAM ID	49	50	51	52	53	55	56	57	58	59	C1	C2	C3
ASPEN ID	49	H2PRD	S10	S28	H2GT	CATOUT	S11	S35	N845	TOCO2CP	CO2PROD	14	15
Mass Flow lb/hr	7673153	40727	40727	40727	40727	797793	797793	797793	152251	600015	593346	593346	593346
Temperature F	1135.9	1391.2	300	85	324.6	1868.9	275	80	80	80	268.3	85	103.6
Pressure psi	954	20.5	19.6	18.5	350	19.5	18.7	14.8	14.8	14.8	2100	2060	3000
H MMBtu/hr	-25500.7	-2881.1	28.6	0.2	32	-2881.4	-3297.1	-3504.2	-1044.1	-2286.6	-2245.8	-2310.7	-2307.4

STREAM ID	C4	H1	H2	H3	H4	H5	H6	H7	T1	T2	T3
ASPEN ID	C4	HVAIR	35	43	12	AIRASU	32	38	31	GTPCX	34
Mass Flow lb/hr	158920	5026390	5026390	5.61E+09	263554	1074908	3899243	3899243	3939979	4203533	4203533
Temperature F	80.9	59	66	59	100	60	60	1050	2585.2	1115.9	246
Pressure psi	14.8	14.7	15.3	58.6	120	282	282	282	268.5	15	15
H MMBtu/hr	-1089.5	-209.9	-201.4	-3.84E+07	-11.8	-53.8	-195	810.6	815.8	-934.1	-1939.7

HRSG/STEAM CYCLE

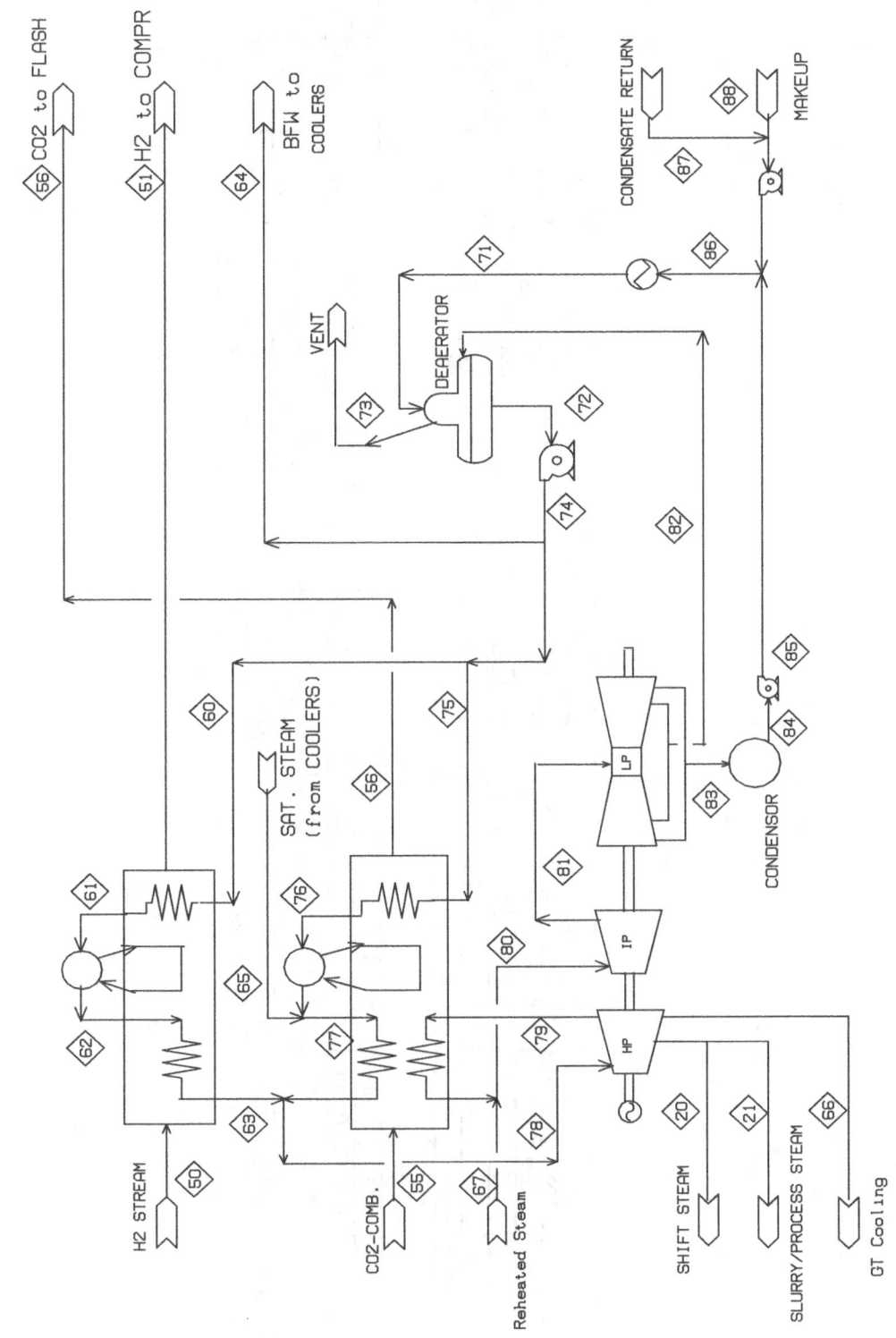

A-61

HYDRAULIC AIR COMPRESSION CYCLE - COAL SYNGAS - CO2 SEQUESTRATION

HRSG / STEAM CYCLE

STREAM ID	20	21	60	61	62	63	64	65	66	67	71	72	73	74
ASPEN ID	SHFSTM	SLURSTM	STMCH4	S22	S23	S25	OSYNCO	RSYNCO	45	51	TODEAER	TOPMPHF	DVENT	S17
Mass Flow lb/hr	272791	79025	107747	107747	106670	106670	208156	208156	80000	80000	554276	560708	2818	560708
Temperature F	875	879.1	221.3	620	629.3	1050	221.3	635	709.8	1100.9	205	217.3	217.3	221.3
Pressure psi	1000	1000	2345.6	2011.1	1910.5	1800	2345.6	1911	518	492.1	17.1	16.3	16.3	2345.6
H MMBtu/hr	-1482.8	-429.1	-718.8	-670.3	-610.3	-571.2	-1388.7	-1188.5	-440.3	-423.3	-3709.8	-3745.9	-16.1	-3740.8

STREAM ID	75	76	77	78	79	80	81	82	83	84	85	86	87	88
ASPEN ID	TSTMCO2	TOBLR	S19	S20	44	IPTURIN	IPTUREXL	LPDEAER	VLPEX	CNDOUT	TOMIX	TOCNDQ	SLURCND	MKUP
Mass Flow lb/hr	244805	244805	450513	557182	125367	205367	205367	9249	196117	196117	196117	554276	79025	279134
Temperature F	221.3	620	631.8	1050	709.8	1069.8	570.5	355	92.3	91	91	98.2	180	80
Pressure psi	2345.6	2011.1	1910.5	1800	518	492.1	63	17.1	0.8	0.7	20	20	20	20
H MMBtu/hr	-1633.2	-1523	-2575.1	-2983.8	-690.1	-1090.1	-1139.6	-52.2	-1145.8	-1335	-1335	-3769	-530.9	-1903.1

Rocket Engine (CES) - CO$_2$ Capture

Natural Gas CES (gas generator)

CES - Natural Gas - 400 MWe

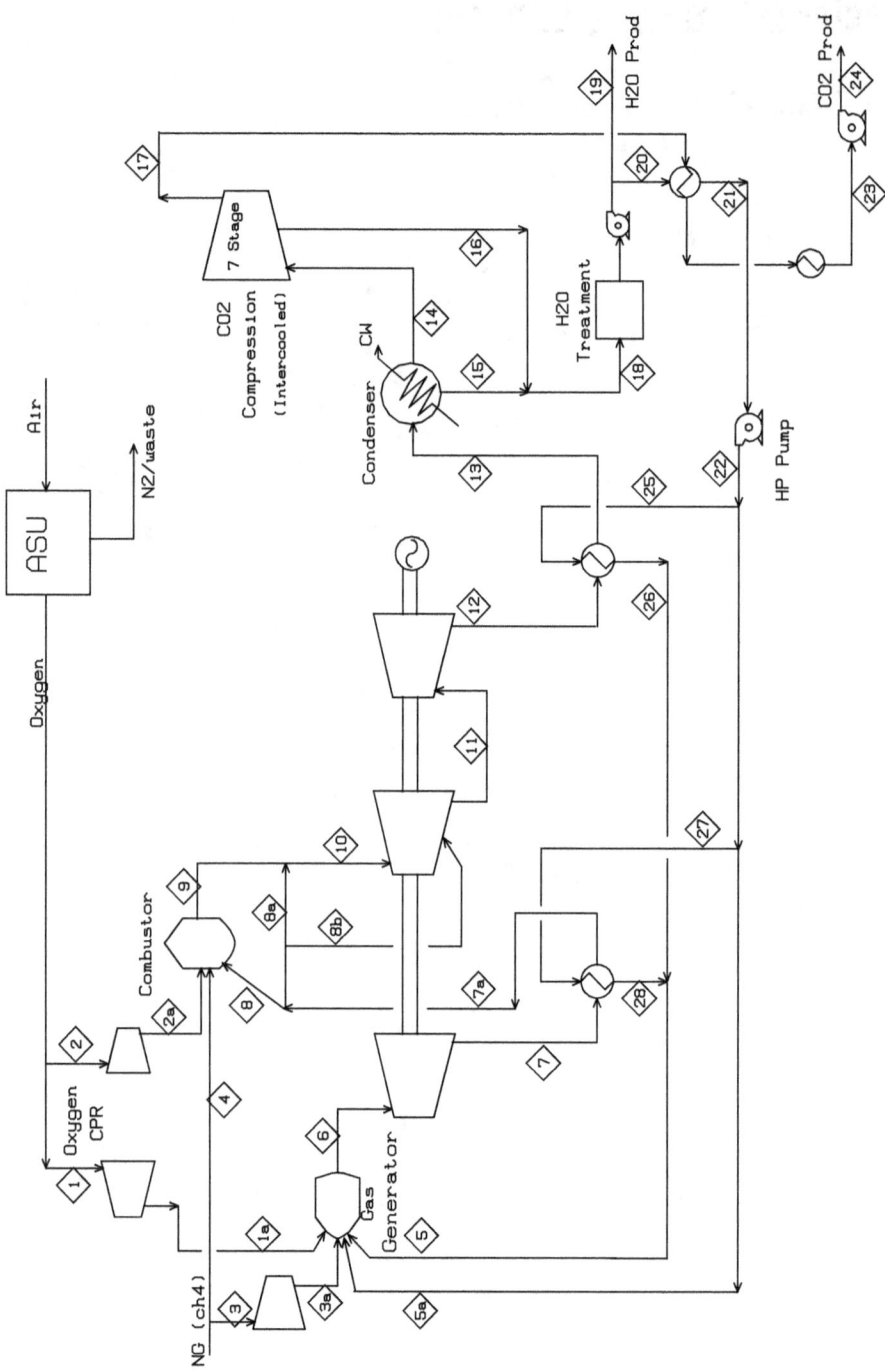

CES - NETL System

A-64

Stream Results Summary   400 MWe - Natural Gas Case

Stream ID	1	1A	2	2A	3	3A	4	5	5A	6	7	7A	8
Temperature F	90.0	264.0	90.0	300.2	90.0	253.0	90.0	674.0	125.0	1850.5	1279.3	600.0	600.0
Pressure psi	30.0	2500.0	420.0	420.0	420.0	2500.0	420.0	2500.0	2600.0	2150.0	400.0	390.0	390.0
Mass Flow lb/hr	210594.0	210594.0	319668.0	319668.0	52000.0	52000.0	78933.0	864488.0	126976.0	1254060.0	1254060.0	1254060.0	1141194.0
Mass Flow lb/sec	58.5	58.5	88.8	88.8	14.4	14.4	21.9	240.1	35.3	348.3	348.3	348.3	317.0
Mole Flow lbmol/hr	6580.5	6580.5	9988.8	9988.8	3241.4	3241.4	4920.2	47985.8	7048.1	64855.8	64855.8	64855.8	59018.8
Enthalpy MMBtu/hr	0.6	7.3	0.8	15.5	-104.2	-100.6	-158.1	-4928.5	-859.3	-5881.1	-6257.6	-6685.5	-6083.8
Vapor Frac	1.000	1.000	1.000	1.000	1.000	1.000	1.000	1.000	0.000	1.000	1.000	1.000	1.000
Cp Btu/lb-R	0.220	0.260	0.220	0.232	0.581	0.734	0.581	1.863	1.004	0.590	0.524	0.489	0.489

Mass Flow lb/hr

	1	1A	2	2A	3	3A	4	5	5A	6	7	7A	8
O2	209515.5	209515.5	318030.7	318030.7	0.0	0.0	0.0	0.0	0.0	2074.4	2074.4	2074.4	1887.7
N2	553.0	553.0	839.5	839.5	0.0	0.0	0.0	0.0	0.0	553.0	553.0	553.0	503.3
AR	525.8	525.8	798.1	798.1	0.0	0.0	0.0	0.0	0.0	525.8	525.8	525.8	478.4
H2	0.0	0.0	0.0	0.0	0.0	0.0	0.0	0.0	0.0	0.0	0.0	0.0	0.0
CO	0.0	0.0	0.0	0.0	0.0	0.0	0.0	0.0	0.0	0.0	0.0	0.0	0.0
CO2	0.0	0.0	0.0	0.0	0.0	0.0	0.0	0.0	0.0	142652.0	142652.0	142652.0	129813.3
H2O	0.0	0.0	0.0	0.0	0.0	0.0	0.0	864488.4	126975.7	1108260.0	1108260.0	1108260.0	1008510.0
CH4	0.0	0.0	0.0	0.0	52000.0	52000.0	78932.5	0.0	0.0	0.0	0.0	0.0	0.0

Mole Flow lbmol/hr

	1	1A	2	2A	3	3A	4	5	5A	6	7	7A	8
O2	6547.6	6547.6	9938.8	9938.8	0.0	0.0	0.0	0.0	0.0	64.8	64.8	64.8	59.0
N2	19.7	19.7	30.0	30.0	0.0	0.0	0.0	0.0	0.0	19.7	19.7	19.7	18.0
AR	13.2	13.2	20.0	20.0	0.0	0.0	0.0	0.0	0.0	13.2	13.2	13.2	12.0
H2	0.0	0.0	0.0	0.0	0.0	0.0	0.0	0.0	0.0	0.0	0.0	0.0	0.0
CO	0.0	0.0	0.0	0.0	0.0	0.0	0.0	0.0	0.0	0.0	0.0	0.0	0.0
CO2	0.0	0.0	0.0	0.0	0.0	0.0	0.0	0.0	0.0	3241.4	3241.4	3241.4	2949.7
H2O	0.0	0.0	0.0	0.0	0.0	0.0	0.0	47985.8	7048.1	61516.7	61516.7	61516.7	55980.2
CH4	0.0	0.0	0.0	0.0	3241.4	3241.4	4920.2	0.0	0.0	0.0	0.0	0.0	0.0

Mole Frac

	1	1A	2	2A	3	3A	4	5	5A	6	7	7A	8
O2	0.995	0.995	0.995	0.995	0.000	0.000	0.000	0.000	0.000	0.001	0.001	0.001	0.001
N2	0.003	0.003	0.003	0.003	0.000	0.000	0.000	0.000	0.000	0.000	0.000	0.000	0.000
AR	0.002	0.002	0.002	0.002	0.000	0.000	0.000	0.000	0.000	0.000	0.000	0.000	0.000
H2	0.000	0.000	0.000	0.000	0.000	0.000	0.000	0.000	0.000	0.000	0.000	0.000	0.000
CO	0.000	0.000	0.000	0.000	0.000	0.000	0.000	0.000	0.000	0.000	0.000	0.000	0.000
CO2	0.000	0.000	0.000	0.000	0.000	0.000	0.000	0.000	0.000	0.050	0.050	0.050	0.050
H2O	0.000	0.000	0.000	0.000	0.000	0.000	0.000	1.000	1.000	0.949	0.949	0.949	0.949
CH4	0.000	0.000	0.000	0.000	1.000	1.000	1.000	0.000	0.000	0.000	0.000	0.000	0.000
total	1.0	1.0	1.0	1.0	1.0	1.0	1.0	1.0	1.0	1.0	1.0	1.0	1.0

Stream Results Summary   400 MWe - Natural Gas Case

Stream ID	8A	8B	8C	9	10	11	12	13	14	15	16	17	18
Temperature F	600.0	600.0	600.0	2667.5	2600.0	1382.9	791.5	139.0	100.0	100.0	100.0	245.4	100.0
Pressure psi	390.0	390.0	390.0	380.0	380.0	18.1	2.1	2.0	1.9	1.9	5.7	2100.0	1.9
Mass Flow lb/hr	37622.0	60195.0	15049.0	1539797.0	1577419.0	1652662.0	1652662.0	1652662.0	519206.0	1133456.0	151077.0	368089.0	1284533.0
Mass Flow lb/sec	10.5	16.7	4.2	427.7	438.2	459.1	459.1	459.1	144.2	314.8	42.0	102.2	356.8
Mole Flow lbmol/hr	1945.7	3113.1	778.3	73927.8	75873.5	79764.8	79764.8	79764.8	16849.7	62915.1	8385.3	8463.2	71300.4
Enthalpy MMBtu/hr	-200.6	-320.9	-80.2	-6226.4	-6445.8	-7848.1	-8306.8	-8759.2	-2255.2	-7707.8	-1027.3	-1387.7	-8735.1
Vapor Frac	1.000	1.000	1.000	1.000	1.000	1.000	1.000	1.000	1.000	0.000	0.000	1.000	0.000
Cp Btu/lb-R	0.489	0.489	0.489	0.579	0.576	0.494	0.445	0.395	0.276	1.017	1.017	0.398	1.017
Mass Flow lb/hr													
O2	62.2	99.6	24.9	5036.5	5098.7	5223.2	5223.2	5223.2	5223.2	0.0	0.0	5197.1	0.0
N2	16.6	26.5	6.6	1342.7	1359.3	1392.5	1392.5	1392.5	1392.5	0.0	0.0	1385.5	0.0
AR	15.8	25.2	6.3	1276.5	1292.3	1323.8	1323.8	1323.8	1323.8	0.0	0.0	1317.2	0.0
H2	0.0	0.0	0.0	0.0	0.0	0.0	0.0	0.0	0.0	0.0	0.0	0.0	0.0
CO	0.0	0.0	0.0	0.0	0.0	0.0	0.0	0.0	0.0	0.0	0.0	0.0	0.0
CO2	4279.6	6847.3	1711.8	346349.4	350629.0	359188.1	359188.1	359188.1	359172.9	15.2	21.4	359151.5	36.5
H2O	33247.6	53196.2	13299.1	1185790.0	1219040.0	1285540.0	1285540.0	1285540.0	152093.3	1133440.0	151055.9	1038.0	1284500.0
CH4	0.0	0.0	0.0	0.0	0.0	0.0	0.0	0.0	0.0	0.0	0.0	0.0	0.0
Mole Flow lbmol/hr													
O2	1.9	3.1	0.8	157.4	159.3	163.2	163.2	163.2	163.2	0.0	0.0	162.4	0.0
N2	0.6	0.9	0.2	47.9	48.5	49.7	49.7	49.7	49.7	0.0	0.0	49.5	0.0
AR	0.4	0.6	0.2	32.0	32.3	33.1	33.1	33.1	33.1	0.0	0.0	33.0	0.0
H2	0.0	0.0	0.0	0.0	0.0	0.0	0.0	0.0	0.0	0.0	0.0	0.0	0.0
CO	0.0	0.0	0.0	0.0	0.0	0.0	0.0	0.0	0.0	0.0	0.0	0.0	0.0
CO2	97.2	155.6	38.9	7869.9	7967.1	8161.6	8161.6	8161.6	8161.3	0.3	0.5	8160.8	0.8
H2O	1845.5	2952.8	738.2	65820.6	67666.2	71357.2	71357.2	71357.2	8442.4	62914.8	8384.8	57.6	71299.5
CH4	0.0	0.0	0.0	0.0	0.0	0.0	0.0	0.0	0.0	0.0	0.0	0.0	0.0
Mole Frac													
O2	0.001	0.001	0.001	0.002	0.002	0.002	0.002	0.002	0.010	0.000	0.000	0.019	0.000
N2	0.000	0.000	0.000	0.001	0.001	0.001	0.001	0.001	0.003	0.000	0.000	0.006	0.000
AR	0.000	0.000	0.000	0.000	0.000	0.000	0.000	0.000	0.002	0.000	0.000	0.004	0.000
H2	0.000	0.000	0.000	0.000	0.000	0.000	0.000	0.000	0.000	0.000	0.000	0.000	0.000
CO	0.000	0.000	0.000	0.000	0.000	0.000	0.000	0.000	0.000	0.000	0.000	0.000	0.000
CO2	0.050	0.050	0.050	0.106	0.105	0.102	0.102	0.102	0.484	0.000	0.000	0.964	0.000
H2O	0.949	0.949	0.949	0.890	0.892	0.895	0.895	0.895	0.501	1.000	1.000	0.007	1.000
CH4	0.000	0.000	0.000	0.000	0.000	0.000	0.000	0.000	0.000	0.000	0.000	0.000	0.000
total	1.0	1.0	1.0	1.0	1.0	1.0	1.0	1.0	1.0	1.0	1.0	1.0	1.0

Stream Results Summary 400 MWe - Natural Gas Case

Stream ID		19	20	21	21A	22	23	24	25	26	27	28
Temperature	F	100.0	100.0	125.0	100.0	127.6	100.0	123	127.6	675.7	127.6	675.6
Pressure	psi	50.0	50.0	47.5	1.9	2600.0	2060.0	3000	2600	2500	2600	2500
Mass Flow	lb/hr	293033.0	991464.0	991464.0	1284497.0	991464.0	368089.0	368089	444258	444258	420230	420230
Mass Flow	lb/sec	81.4	275.4	275.4	356.8	275.4	102.2	102.2	123.4	123.4	116.7	116.7
Mole Flow	lbmol/hr	16265.6	55033.9	55033.9	71299.5	55033.9	8463.2	8463.2	24659.8	24659.8	23326.1	23326.1
Enthalpy	MMBtu/hr	-1992.0	-6739.8	-6715.1	-8735.0	-6706.2	-1420.8	-1418.7	-3004.9	-2552.5	-2842.4	-2414.5
Vapor Frac		0.000	0.000	0.000	0.000	0.000	0.000	0.000	0.000	1.000	0.000	1.000
Cp	Btu/lb-R	0.996	0.996	0.997	1.017	0.988	0.706	0.578	0.988	2.905	0.988	2.913

Mass Flow lb/hr

	19	20	21	21A	22	23	24	25	26	27	28
O2	0.0	0.0	0.0	0.0	0.0	5197.1	5197.1	0.0	0.0	0.0	0.0
N2	0.0	0.0	0.0	0.0	0.0	1385.5	1385.5	0.0	0.0	0.0	0.0
AR	0.0	0.0	0.0	0.0	0.0	1317.2	1317.2	0.0	0.0	0.0	0.0
H2	0.0	0.0	0.0	0.0	0.0	0.0	0.0	0.0	0.0	0.0	0.0
CO	0.0	0.0	0.0	0.0	0.0	0.0	0.0	0.0	0.0	0.0	0.0
CO2	0.0	0.0	0.0	0.0	0.0	359151.5	359151.5	0.0	0.0	0.0	0.0
H2O	293032.7	991464.1	991464.1	1284500.0	991464.1	1038.0	1038.0	444257.9	444257.9	420230.5	420230.5
CH4	0.0	0.0	0.0	0.0	0.0	0.0	0.0	0.0	0.0	0.0	0.0

Mole Flow lbmol/hr

	19	20	21	21A	22	23	24	25	26	27	28
O2	0.0	0.0	0.0	0.0	0.0	162.4	162.4	0.0	0.0	0.0	0.0
N2	0.0	0.0	0.0	0.0	0.0	49.5	49.5	0.0	0.0	0.0	0.0
AR	0.0	0.0	0.0	0.0	0.0	33.0	33.0	0.0	0.0	0.0	0.0
H2	0.0	0.0	0.0	0.0	0.0	0.0	0.0	0.0	0.0	0.0	0.0
CO	0.0	0.0	0.0	0.0	0.0	0.0	0.0	0.0	0.0	0.0	0.0
CO2	0.0	0.0	0.0	0.0	0.0	8160.8	8160.8	0.0	0.0	0.0	0.0
H2O	16265.6	55033.9	55033.9	71299.5	55033.9	57.6	57.6	24659.8	24659.8	23326.1	23326.1
CH4	0.0	0.0	0.0	0.0	0.0	0.0	0.0	0.0	0.0	0.0	0.0

Mole Frac

	19	20	21	21A	22	23	24	25	26	27	28
O2	0.000	0.000	0.000	0.000	0.000	0.019	0.019	0.000	0.000	0.000	0.000
N2	0.000	0.000	0.000	0.000	0.000	0.006	0.006	0.000	0.000	0.000	0.000
AR	0.000	0.000	0.000	0.000	0.000	0.004	0.004	0.000	0.000	0.000	0.000
H2	0.000	0.000	0.000	0.000	0.000	0.000	0.000	0.000	0.000	0.000	0.000
CO	0.000	0.000	0.000	0.000	0.000	0.000	0.000	0.000	0.000	0.000	0.000
CO2	0.000	0.000	0.000	0.000	0.000	0.964	0.964	0.000	0.000	0.000	0.000
H2O	1.000	1.000	1.000	1.000	1.000	0.007	0.007	1.000	1.000	1.000	1.000
CH4	0.000	0.000	0.000	0.000	0.000	0.000	0.000	0.000	0.000	0.000	0.000
total	1.0	1.0	1.0	1.0	1.0	1.0	1.0	1.0	1.0	1.0	1.0

Natural Gas CES (gas generator)

POWER SUMMARY (CH4 FUEL)

POWER kW

CO2 Compression

CO2 Compressor #1	13428.52	
CO2 Compressor #2	8573.21	
CO2 Compressor #3	8147.99	
total		30149.71

O2 Plant/Compressors

Oxygen Plant	52933.98	
HP O2 Compressor	13721.52	
IP O2 Compressor	12513.97	
total		79169.47

Fuel Compressor 2425.37

Pumps/Fans

Condensate Pump	63.58	
HP H2O Recycle Pump	2619.42	
HP CO2 Pump	610.15	
Water Pumps	8069.02	
Cooling Tower Fans	2570.15	
		13932.33

Turbine Power

HP Turbine	-108696.35	
IP Turbine	-289003.18	
LP Turbine	-132418.05	
Total Turbines		-530117.58

(with CO2 Sequestration)

Gross Power	-404440.70
Auxiliary (1.5%)	6066.61
Net Power	-398374.09

Efficiency

% LHV	48.27
% HHV	43.63

(WITHOUT CO2 SEQUESTRATION)

Gross Power	-421161.89
Auxiliary (1.5%)	6317.43
Net Power	-414844.47

Efficiency

% LHV	50.26820776
% HHV	45.43182732

(with CO2 Sequestration)	
Gross Power	-404440.70
Auxiliary (1.5%)	6066.61
Net Power (KWe)	-398374.09
Efficiency	
% LHV	48.27
% HHV	43.63

(without CO2 SEQUESTRATION)	
Gross Power	-421161.89
Auxiliary (1.5%)	6317.43
Net Power	-414844.47
Efficiency	
% LHV	50.27
% HHV	45.43

Rocket Engine (CES) - CO$_2$ Capture

Coal Syngas CES (gas generator) – Destec HP / HGCU

Destec Gasification / CES Power Generation / CO2 Sequestration (406 MWe)

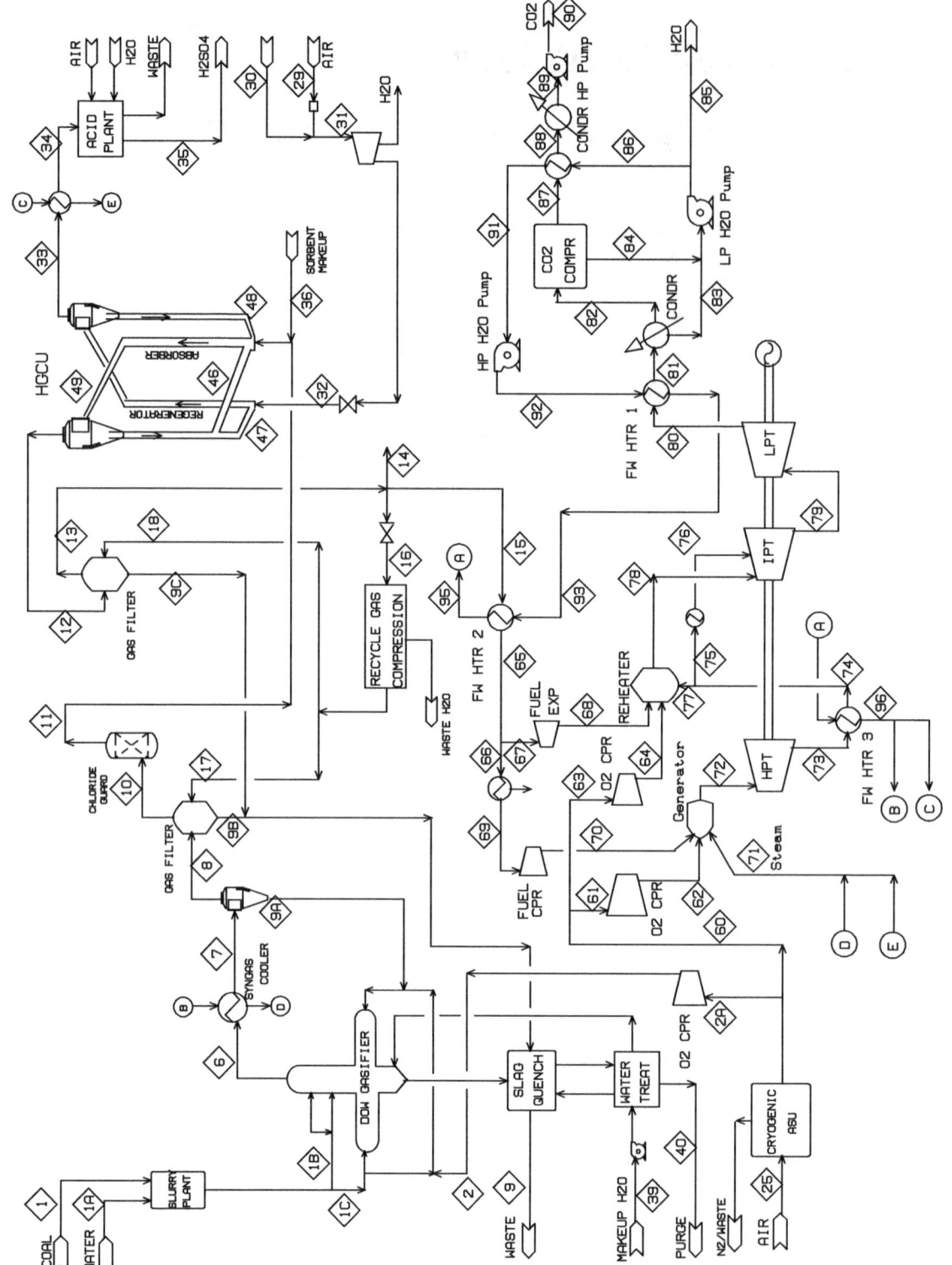

Destec Gasification - CES Power Generation
(mass/energy balances)

PFD STREAM #	1	1A	1B	1C	2A	2	6	7	8	9A	9B	9C	9	39	40
ASPEN NAME	COLIN	WAT1	COLB	COLA	GO2A	GOXYG	DRXROUT	RAWPRD	DRAWGAZ	FNES	16	19	WSTSOL	MWATG	PURGE
Temperature F	100	100	350	350	60	289.2	1905	1110	1110	1110	1099.4	1127.4	200	59	200
Pressure psi	14.7	14.7	1078	1078	18	1150	1034	1024	1019	1019	14.7	14.7	14.7	15	15
Mass Flow lb/hr	297507	99131	82970	313666	198509	198509	577138	577138	567090	10047	1061	553	44780	36051	20943
Mass Flow lb/sec	82.6	27.5	23	87.1	55.1	55.1	160.3	160.3	157.5	2.8	0.3	0.2	12.4	10	5.8
Mole Flow lbmol/hr		5502.5			6195.9	6195.9								2001.1	1162.5
Enthalpy MMBtu/hr	-920.2	-681.1	-185	-692.8	-0.8	8.6	-919.3	-1123.4	-1110.5	-12.9	-1.4	-2.1	-166.7	-249.4	-141.5

PFD STREAM #	11	12	13	14	15	16	17	18	25	29	30	33	34	35	46
ASPEN NAME	18	20	26	FUELASU	TOCES	21	25	24C	AIRTOT	RAIR	30X	5A	5	SACID	46
Temperature F	1095.5	1133.3	1127.4	1127.4	1127.4	1127.4	334.2	334.2	59	59	80	1382.7	850	100	1132.2
Pressure psi	985	975	965	964	964	965	1291.22	1291.22	14.7	14.55	14.8	955	940	16	975
Mass Flow lb/hr	573269	570035	573857	1511	559642	12712	8112	4375	2682809	46679	50228	99872	99872	22990	7105557
Mass Flow lb/sec	159.2	158.3	159.4	0.4	155.5	3.5	2.3	1.2	745.2	13	14	27.7	27.7	6.4	1973.8
Mole Flow lbmol/hr			30447.7	80.1	29693.5	674.5	430.1	231.9	92971.1	1617.6	1164.4	2623.4	2623.4	236.9	61072.5
Enthalpy MMBtu/hr	-1127.8	-1142.9	-1150.8	-3	-1122.3	-25.5	-18.5	-10	-112	-1.9	-191.3	-185.8	-200.2	-28.8	-24298.5

PFD STREAM #	48	49	60	61	62	63	64	65	66	67	68	69	70	71	72
ASPEN NAME	48	49	O260S	O2GEN	O2GENX	O2IPT	O2MEDX	50	FUELHPT	FUELHPT	41	HPFTCPR	FUELHPX	INJMIX	TOHP
Temperature F	1382.7	1133.3	60	60	284.8	60	277.8	680	680	680	518.6	205.4	202.8	798.2	1850
Pressure psi	955	954	18	18	2500	18	420	935	935	935	41	907	2500	2500	2150
Mass Flow lb/hr	785775	8465098	393061	128972	128972	264089	264089	559642	183638	376026	376026	171668	169030	1059534	1357535
Mass Flow lb/sec	218.3	2351.4	109.2	35.8	35.8	73.4	73.4	155.5	51	104.5	104.5	47.7	47	294.3	377.1
Mole Flow lbmol/hr	6553.6		12268.4	4025.5	4025.5	8242.9	8242.9	29693.5	9748.6	19961.8	19961.8	9084.2	8937.9	58812.4	67823.5
Enthalpy MMBtu/hr	-2701.3	-28139.3	-1.5	-0.5	5.1	-1	11.4	-1229	-403.3	-825.8	-850.9	-370.4	-355.9	-5896.6	-6256.7

PFD STREAM #	74	75	76	77	78	79	80	81	82	83	84	85	86	87	88
ASPEN NAME	25X	35	36	TORHT	RHT1EX	TOLP	TOHTREC	POC5	34	33	PINJWAT	H2OPROD	INJH2O	TOREF	37
Temperature F	680	680	620	680	2599.3	1416	828	143	34	100	100	100	244.9	244.9	158.8
Pressure psi	390	390	380	390	380	18.1	2.1	2.1	1.9	1.9	5.62	50	50	2100	2100
Mass Flow lb/hr	1357535	95027	95027	1262508	1934298	1997649	1997649	1997649	992970	1004679	290457	235451	1059534	702444	702444
Mass Flow lb/sec	377.1	26.4	26.4	350.7	537.3	554.9	554.9	554.9	275.8	279.1	80.7	65.4	294.3	195.1	195.1
Mole Flow lbmol/hr	67823.5	4747.6	4747.6	63075.8	84769.9	87935	87935	87935	32168.7	55766.3	16118.7	13069.4	58812.4	58812.4	16048.2
Enthalpy MMBtu/hr	-7052.9	-493.7	-496.4	-6559.2	-7583.4	-9034.2	-9552	-10088	-4331.7	-6832	-1974.6	-1600.6	-7202.5	-2664.8	-2696.4

PFD STREAM #	90	91	92	93	95	96
ASPEN NAME	CO2PROD	38	26X	53	40	11
Temperature F	122.5	130	133.3	602.8	663.8	674.8
Pressure psi	3000	25	2885	2797	2713	2577.3
Mass Flow lb/hr	702444	1059534	1059534	1059534	1059534	1059534
Mass Flow lb/sec	195.1	294.3	294.3	294.3	294.3	294.3
Mole Flow lbmol/hr	16048.2	58812.4	58812.4	58812.4	58812.4	58812.4
Enthalpy MMBtu/hr	-2724.2	-7171	-7159.9	-6623.9	-6517.1	-6115.4

CES Process Streams

	Fuel	Oxygen Streams					Fuel Streams						Steam-generator
PFD STREAM #	15	60	61	62	63	64	65	66	67	68	69	70	71
ASPEN NAME	TOCES	O260S	O2GEN	O2GENX	O2IPT	O2MEDX	50	FUELHPT	FUELIPT	41	HPFTCPR	FUELHPX	INJMIX
Temperature F	1127.4	60.0	60.0	284.8	60.0	277.8	680.0	680.0	680.0	518.6	205.4	202.8	798.2
Pressure psi	964.0	18.0	18.0	2500.0	18.0	420.0	935.0	935.0	935.0	475.0	907.0	2500.0	2500.0
Mass Flow lb/hr	559642	393061	128972	128972	264089	264089	559642	183638	376026	376026	171668	169030	1059534
Mass Flow lb/sec	155.5	109.2	35.8	35.8	73.4	73.4	155.5	51.0	104.5	104.5	47.7	47.0	294.3
Mole Flow lbmol/hr	29693.5	12268.4	4025.0	4025.5	8242.9	8242.9	29693.5	9748.6	19961.8	19961.8	9084.2	8937.9	58812.4
Enthalpy MMBtu/hr	-1122.3	-1.5	-0.5	5.1	-1.0	11.4	-1229.0	-403.3	-825.8	-850.9	-370.4	-355.9	-5896.6
Average MW	18.847	32.039	32.039	32.039	32.039	32.039	18.847	18.837	18.837	18.837	18.897	18.912	18.016
CPMX Btu/lb-R	0.435	0.219	0.219	0.258	0.219	0.231	0.417	0.417	0.417	0.409	0.406	0.430	0.831
Mole Frac													
O2	0.0000	0.9950	0.9950	0.9950	0.9950	0.9950	0.0000	0.0000	0.0000	0.0000	0.0000	0.0000	0.0000
N2	0.0036	0.0000	0.0000	0.0000	0.0000	0.0000	0.0036	0.0038	0.0038	0.0038	0.0041	0.0041	0.0000
AR	0.0010	0.0050	0.0050	0.0050	0.0050	0.0050	0.0010	0.0010	0.0010	0.0010	0.0011	0.0011	0.0000
H2	0.3803	0.0000	0.0000	0.0000	0.0000	0.0000	0.3803	0.3808	0.3808	0.3808	0.4086	0.4153	0.0000
CO	0.4295	0.0000	0.0000	0.0000	0.0000	0.0000	0.4295	0.4294	0.4294	0.4294	0.4608	0.4684	0.0000
CO2	0.0988	0.0000	0.0000	0.0000	0.0000	0.0000	0.0988	0.0986	0.0986	0.0986	0.1059	0.1076	0.0000
H2O	0.0840	0.0000	0.0000	0.0000	0.0000	0.0000	0.0840	0.0841	0.0841	0.0841	0.0172	0.0011	1.0000
CH4	0.0011	0.0000	0.0000	0.0000	0.0000	0.0000	0.0011	0.0009	0.0009	0.0009	0.0009	0.0009	0.0000
H2S	0.0000	0.0000	0.0000	0.0000	0.0000	0.0000	0.0000	0.0000	0.0000	0.0000	0.0000	0.0000	0.0000
SO2	0.0000	0.0000	0.0000	0.0000	0.0000	0.0000	0.0000	0.0000	0.0000	0.0000	0.0000	0.0000	0.0000
CL2	0.0000	0.0000	0.0000	0.0000	0.0000	0.0000	0.0000	0.0000	0.0000	0.0000	0.0000	0.0000	0.0000
HCL	0.0000	0.0000	0.0000	0.0000	0.0000	0.0000	0.0000	0.0000	0.0000	0.0000	0.0000	0.0000	0.0000
NH3	0.0017	0.0000	0.0000	0.0000	0.0000	0.0000	0.0017	0.0014	0.0014	0.0014	0.0015	0.0015	0.0000
COS	0.0000	0.0000	0.0000	0.0000	0.0000	0.0000	0.0000	0.0000	0.0000	0.0000	0.0000	0.0000	0.0000
NO2	0.0000	0.0000	0.0000	0.0000	0.0000	0.0000	0.0000	0.0000	0.0000	0.0000	0.0000	0.0000	0.0000
Mass Frac													
O2	0.0000	0.9938	0.9938	0.9938	0.9938	0.9938	0.0000	0.0000	0.0000	0.0000	0.0000	0.0000	0.0000
N2	0.0053	0.0000	0.0000	0.0000	0.0000	0.0000	0.0053	0.0056	0.0056	0.0056	0.0060	0.0061	0.0000
AR	0.0022	0.0062	0.0062	0.0062	0.0062	0.0062	0.0022	0.0022	0.0022	0.0022	0.0024	0.0024	0.0000
H2	0.0407	0.0000	0.0000	0.0000	0.0000	0.0000	0.0407	0.0407	0.0407	0.0407	0.0436	0.0443	0.0000
CO	0.6384	0.0000	0.0000	0.0000	0.0000	0.0000	0.6384	0.6385	0.6385	0.6385	0.6830	0.6937	0.0000
CO2	0.2307	0.0000	0.0000	0.0000	0.0000	0.0000	0.2307	0.2305	0.2305	0.2305	0.2465	0.2504	0.0000
H2O	0.0802	0.0000	0.0000	0.0000	0.0000	0.0000	0.0802	0.0805	0.0805	0.0805	0.0164	0.0010	1.0000
CH4	0.0009	0.0000	0.0000	0.0000	0.0000	0.0000	0.0009	0.0007	0.0007	0.0007	0.0008	0.0008	0.0000
H2S	0.0000	0.0000	0.0000	0.0000	0.0000	0.0000	0.0000	0.0000	0.0000	0.0000	0.0000	0.0000	0.0000
SO2	0.0000	0.0000	0.0000	0.0000	0.0000	0.0000	0.0000	0.0000	0.0000	0.0000	0.0000	0.0000	0.0000
CL2	0.0000	0.0000	0.0000	0.0000	0.0000	0.0000	0.0000	0.0000	0.0000	0.0000	0.0000	0.0000	0.0000
HCL	0.0000	0.0000	0.0000	0.0000	0.0000	0.0000	0.0000	0.0000	0.0000	0.0000	0.0000	0.0000	0.0000
NH3	0.0016	0.0000	0.0000	0.0000	0.0000	0.0000	0.0016	0.0012	0.0012	0.0012	0.0013	0.0013	0.0000
COS	0.0000	0.0000	0.0000	0.0000	0.0000	0.0000	0.0000	0.0000	0.0000	0.0000	0.0000	0.0000	0.0000
NO2	0.0000	0.0000	0.0000	0.0000	0.0000	0.0000	0.0000	0.0000	0.0000	0.0000	0.0000	0.0000	0.0000

A-72

CES Process Streams

	To HP Turbine	From HP turbine	From FW Heater #3	Turbine cooling		Gas - Reheater	To IP Turbine	To LP Turbine	From LP Turbine	From FW Heater #1	To CO2 Compr	H2O - Condenser	H2O - CO2 CPR
PFD STREAM #	72	73	74	75	76	77	78	79	80	81	82	83	84
ASPEN NAME	TOHP	TOIP	25X	35	36	TORHT	RHT1EX	TOLP	TOHTREC	POC5	34	33	PINJWAT
Temperature F	1850.0	1283.1	680.0	680.0	620.0	680.0	2599.3	1416.0	828.0	143.0	100.0	100.0	100.0
Pressure psi	2150.0	400.0	390.0	390.0	380.0	390.0	380.0	18.1	2.1	2.0	1.9	1.9	5.6
Mass Flow lb/hr	1357535	1357535	1357535	95027	95027	1262508	1934298	1997649	1997649	1997649	992970	1004679	290457
Mass Flow lb/sec	377.1	377.1	377.1	26.4	26.4	350.7	537.3	554.9	554.9	554.9	275.8	279.1	80.7
Mole Flow lbmol/hr	67823.5	67823.5	67823.5	4747.6	4747.6	63075.8	84769.9	87935.0	87935.0	87935.0	32168.7	55766.3	16118.7
Enthalpy MMBtu/hr	-6256.7	-6651.2	-7052.9	-493.7	-496.4	-6559.2	-7583.4	-9034.2	-9552.0	-10088.0	-4331.7	-6832.0	-1974.6
Average MW	20.016	20.016	20.016	20.016	20.016	20.016	22.818	22.717	22.717	22.717	30.868	18.016	18.020
CPMX Btu/lb-R	0.573	0.510	0.475	0.475	0.474	0.475	0.534	0.464	0.418	0.365	0.276	1.017	1.016
Mole Frac													
O2	0.0002	0.0002	0.0002	0.0002	0.0002	0.0002	0.0006	0.0006	0.0006	0.0006	0.0016	0.0000	0.0000
N2	0.0005	0.0005	0.0005	0.0005	0.0005	0.0005	0.0013	0.0013	0.0013	0.0013	0.0035	0.0000	0.0000
AR	0.0004	0.0004	0.0004	0.0004	0.0004	0.0004	0.0011	0.0010	0.0010	0.0010	0.0029	0.0000	0.0000
H2	0.0000	0.0000	0.0000	0.0000	0.0000	0.0000	0.0000	0.0000	0.0000	0.0000	0.0000	0.0000	0.0000
CO	0.0000	0.0000	0.0000	0.0000	0.0000	0.0000	0.0000	0.0000	0.0000	0.0000	0.0000	0.0000	0.0000
CO2	0.0760	0.0760	0.0760	0.0760	0.0760	0.0760	0.1825	0.1787	0.1787	0.1787	0.4885	0.0000	0.0001
H2O	0.9225	0.9225	0.9225	0.9225	0.9225	0.9225	0.8140	0.8179	0.8179	0.8179	0.5024	1.0000	0.9998
CH4	0.0000	0.0000	0.0000	0.0000	0.0000	0.0000	0.0000	0.0000	0.0000	0.0000	0.0000	0.0000	0.0000
H2S	0.0000	0.0000	0.0000	0.0000	0.0000	0.0000	0.0000	0.0000	0.0000	0.0000	0.0000	0.0000	0.0000
SO2	0.0000	0.0000	0.0000	0.0000	0.0000	0.0000	0.0000	0.0000	0.0000	0.0000	0.0000	0.0000	0.0000
CL2	0.0000	0.0000	0.0000	0.0000	0.0000	0.0000	0.0000	0.0000	0.0000	0.0000	0.0000	0.0000	0.0000
HCL	0.0000	0.0000	0.0000	0.0000	0.0000	0.0000	0.0000	0.0000	0.0000	0.0000	0.0000	0.0000	0.0000
NH3	0.0000	0.0000	0.0000	0.0000	0.0000	0.0000	0.0000	0.0000	0.0000	0.0000	0.0000	0.0000	0.0000
COS	0.0000	0.0000	0.0000	0.0000	0.0000	0.0000	0.0000	0.0000	0.0000	0.0000	0.0000	0.0000	0.0000
NO2	0.0002	0.0002	0.0002	0.0002	0.0002	0.0002	0.0005	0.0005	0.0005	0.0005	0.0012	0.0000	0.0001
Mass Frac													
O2	0.0004	0.0004	0.0004	0.0004	0.0004	0.0004	0.0008	0.0008	0.0008	0.0008	0.0016	0.0000	0.0000
N2	0.0008	0.0008	0.0008	0.0008	0.0008	0.0008	0.0016	0.0016	0.0016	0.0016	0.0032	0.0000	0.0000
AR	0.0009	0.0009	0.0009	0.0009	0.0009	0.0009	0.0019	0.0018	0.0018	0.0018	0.0037	0.0000	0.0000
H2	0.0000	0.0000	0.0000	0.0000	0.0000	0.0000	0.0000	0.0000	0.0000	0.0000	0.0000	0.0000	0.0000
CO	0.0000	0.0000	0.0000	0.0000	0.0000	0.0000	0.0000	0.0000	0.0000	0.0000	0.0000	0.0000	0.0000
CO2	0.1671	0.1671	0.1671	0.1671	0.1671	0.1671	0.3520	0.3462	0.3462	0.3462	0.6964	0.0000	0.0002
H2O	0.8304	0.8304	0.8304	0.8304	0.8304	0.8304	0.6427	0.6487	0.6487	0.6487	0.2932	1.0000	0.9996
CH4	0.0000	0.0000	0.0000	0.0000	0.0000	0.0000	0.0000	0.0000	0.0000	0.0000	0.0000	0.0000	0.0000
H2S	0.0000	0.0000	0.0000	0.0000	0.0000	0.0000	0.0000	0.0000	0.0000	0.0000	0.0000	0.0000	0.0000
SO2	0.0000	0.0000	0.0000	0.0000	0.0000	0.0000	0.0000	0.0000	0.0000	0.0000	0.0000	0.0000	0.0000
CL2	0.0000	0.0000	0.0000	0.0000	0.0000	0.0000	0.0000	0.0000	0.0000	0.0000	0.0000	0.0000	0.0000
HCL	0.0000	0.0000	0.0000	0.0000	0.0000	0.0000	0.0000	0.0000	0.0000	0.0000	0.0000	0.0000	0.0000
NH3	0.0000	0.0000	0.0000	0.0000	0.0000	0.0000	0.0000	0.0000	0.0000	0.0000	0.0000	0.0000	0.0000
COS	0.0000	0.0000	0.0000	0.0000	0.0000	0.0000	0.0000	0.0000	0.0000	0.0000	0.0000	0.0000	0.0000
NO2	0.0004	0.0004	0.0004	0.0004	0.0004	0.0004	0.0009	0.0009	0.0009	0.0009	0.0019	0.0000	0.0002

CES Process Streams

	Excess H2O	Recycle H2O	From CO2 CPR	CO2 - Cooler	CO2 - Liquid	From CO2 Pump		Water to Steam Reheating for Gas Generator			
PFD STREAM #	85	86	87	88	89	90	91	92	93	95	96
ASPEN NAME	H2OPROD	INJH2O	TOREF	37	FRREF	CO2PROD	38	26X	53	40	11
Temperature F	100.0	100.0	244.9	158.8	100.0	122.5	130.0	133.3	602.8	663.8	674.8
Pressure psi	50.0	50.0	2100.0	2100.0	2060.0	3000.0	25.0	2885.0	2797.0	2713.0	2577.3
Mass Flow lb/hr	235451	1059534	702444	702444	702444	702444	1059534	1059534	1059534	1059534	1059534
Mass Flow lb/sec	65.4	294.3	195.1	195.1	195.1	195.1	294.3	294.3	294.3	294.3	294.3
Mole Flow lbmol/hr	13069.4	58812.4	16048.2	16048.2	16048.2	16048.2	58812.4	58812.4	58812.4	58812.4	58812.4
Enthalpy MMBtu/hr	-1600.6	-7202.5	-2664.8	-2696.4	-2728.1	-2724.2	-7171.0	-7159.9	-6623.9	-6517.1	-6115.4
Average MW	18.016	18.016	43.771	43.771	43.771	43.771	18.016	18.016	18.016	18.016	18.016
CPMX Btu/lb-R	0.996	0.996	0.401	0.730	0.693	0.572	0.997	0.987	1.389	2.303	
Mole Frac											
O2	0.0000	0.0000	0.0031	0.0031	0.0031	0.0031	0.0000	0.0000	0.0000	0.0000	0.0000
N2	0.0000	0.0000	0.0070	0.0070	0.0070	0.0070	0.0000	0.0000	0.0000	0.0000	0.0000
AR	0.0000	0.0000	0.0057	0.0057	0.0057	0.0057	0.0000	0.0000	0.0000	0.0000	0.0000
H2	0.0000	0.0000	0.0000	0.0000	0.0000	0.0000	0.0000	0.0000	0.0000	0.0000	0.0000
CO	0.0000	0.0000	0.0000	0.0000	0.0000	0.0000	0.0000	0.0000	0.0000	0.0000	0.0000
CO2	0.0000	0.0000	0.9791	0.9791	0.9791	0.9791	0.0000	0.0000	0.0000	0.0000	0.0000
H2O	1.0000	1.0000	0.0027	0.0027	0.0027	0.0027	1.0000	1.0000	1.0000	1.0000	1.0000
CH4	0.0000	0.0000	0.0000	0.0000	0.0000	0.0000	0.0000	0.0000	0.0000	0.0000	0.0000
H2S	0.0000	0.0000	0.0000	0.0000	0.0000	0.0000	0.0000	0.0000	0.0000	0.0000	0.0000
SO2	0.0000	0.0000	0.0000	0.0000	0.0000	0.0000	0.0000	0.0000	0.0000	0.0000	0.0000
CL2	0.0000	0.0000	0.0000	0.0000	0.0000	0.0000	0.0000	0.0000	0.0000	0.0000	0.0000
HCL	0.0000	0.0000	0.0000	0.0000	0.0000	0.0000	0.0000	0.0000	0.0000	0.0000	0.0000
NH3	0.0000	0.0000	0.0000	0.0000	0.0000	0.0000	0.0000	0.0000	0.0000	0.0000	0.0000
COS	0.0000	0.0000	0.0000	0.0000	0.0000	0.0000	0.0000	0.0000	0.0000	0.0000	0.0000
NO2	0.0000	0.0000	0.0024	0.0024	0.0024	0.0024	0.0000	0.0000	0.0000	0.0000	0.0000
Mass Frac											
O2	0.0000	0.0000	0.0023	0.0023	0.0023	0.0023	0.0000	0.0000	0.0000	0.0000	0.0000
N2	0.0000	0.0000	0.0045	0.0045	0.0045	0.0045	0.0000	0.0000	0.0000	0.0000	0.0000
AR	0.0000	0.0000	0.0052	0.0052	0.0052	0.0052	0.0000	0.0000	0.0000	0.0000	0.0000
H2	0.0000	0.0000	0.0000	0.0000	0.0000	0.0000	0.0000	0.0000	0.0000	0.0000	0.0000
CO	0.0000	0.0000	0.0000	0.0000	0.0000	0.0000	0.0000	0.0000	0.0000	0.0000	0.0000
CO2	0.0000	0.0000	0.9844	0.9844	0.9844	0.9844	0.0000	0.0000	0.0000	0.0000	0.0000
H2O	1.0000	1.0000	0.0011	0.0011	0.0011	0.0011	1.0000	1.0000	1.0000	1.0000	1.0000
CH4	0.0000	0.0000	0.0000	0.0000	0.0000	0.0000	0.0000	0.0000	0.0000	0.0000	0.0000
H2S	0.0000	0.0000	0.0000	0.0000	0.0000	0.0000	0.0000	0.0000	0.0000	0.0000	0.0000
SO2	0.0000	0.0000	0.0000	0.0000	0.0000	0.0000	0.0000	0.0000	0.0000	0.0000	0.0000
CL2	0.0000	0.0000	0.0000	0.0000	0.0000	0.0000	0.0000	0.0000	0.0000	0.0000	0.0000
HCL	0.0000	0.0000	0.0000	0.0000	0.0000	0.0000	0.0000	0.0000	0.0000	0.0000	0.0000
NH3	0.0000	0.0000	0.0000	0.0000	0.0000	0.0000	0.0000	0.0000	0.0000	0.0000	0.0000
COS	0.0000	0.0000	0.0000	0.0000	0.0000	0.0000	0.0000	0.0000	0.0000	0.0000	0.0000
NO2	0.0000	0.0000	0.0025	0.0025	0.0025	0.0025	0.0000	0.0000	0.0000	0.0000	0.0000

POWER SUMMARY

(ASPEN CONVENTION , "+" is power usage, "-" is power generation)

GROSS POWER		-414527
Auxiliary POWER (2% of GROSS POWER)		8291
NET PLANT POWER		-406237
COAL USAGE (lbs/hr , dry)		264424
- HHV (Btu/lb , dry)		13126.00
- LHV (")		12656.94
OVERALL EFFICIENCY		
- HHV basis %		39.96
- LHV basis %		41.44
Thermal Input		
- LHV (KW)	980316.6311	-0.41439496
- HHV (KW)	1016646.846	-0.399585005

CO2 as low pressure gas	(No sequestration - approximate)
Gross Power	-451520.98
Net Power	-442490.56
HHV %	43.52
LHV %	45.14

	POWER kW	
Air Separation Plant		
- Gasification	21835.9557	
- CES (generator+reheater)	43236.6613	
total		65072.617
Oxygen Compression		
- for gasifier	12071.374	
- for CES generator	9380.54256	
- for CES reheater	12066.522	
total		33518.43856
Syngas		
- HP Cpr for CES generator	4460.46151	
- Expander for CES reheater (credit)	-7256.7916	
total		-2796.33009
CO2 Compression		
- #1 (1.9 to 17.85 psia)	25825.9715	
- #2 (17.5 to 163.2 psia)	16320.2041	
- #3 (160 to 2100 psia)	15416.7619	
total		57562.9375
Gasification Misc		
- HGCU/Recycle	5852.94348	
- pumps (slurry, makeup)	228.84847	
total		6081.79195
Cooling tower		
- pumps	6241.94946	
- fan	1978.63286	
total		8220.58232
CES pumps		
- condensate	65.5863712	
- HP water	3254.10419	
- CO2 pump	1146.33903	
total		4466.029591
Power Turbines		
- HP Turb	-113883.77	
- IP Turb	-323285.88	
- LP Turb	-149483.8	
total		-586653.45

CO2 Compression

(ASPEN Representation was a series of three intercooled multistage compressors)

Compressor	# of Stages	Intercooling Temperature ° F	Exit Cooling Temperature ° F	Pressure Inlet (psia)	Pressure Outlet (psia)	Stage Isentropic Efficiency	Stage Mechanical Efficiency	Total Power (KWe)	Gas - Inlet (lbs/sec)	Liquid Prod (lbs/sec)	Total Cooling Duty (MMBtu/Hr)
1	2	100	100	1.9	17.85	0.85	0.985	25826	275.8	76.2	372
2	2	100	100	17.5	163.2	0.85	0.985	16320	199.6	4.1	73
3	3	100	n/a	160	2100	0.85	0.985	15417	195.5	0.0	56
							total: (KWe)	57563			

Hydrogen Turbine - CO$_2$ Capture

Hydrogen from Steam Methane Reforming (SMR)

Hydrogen from Steam Methane Reforming (SMR)

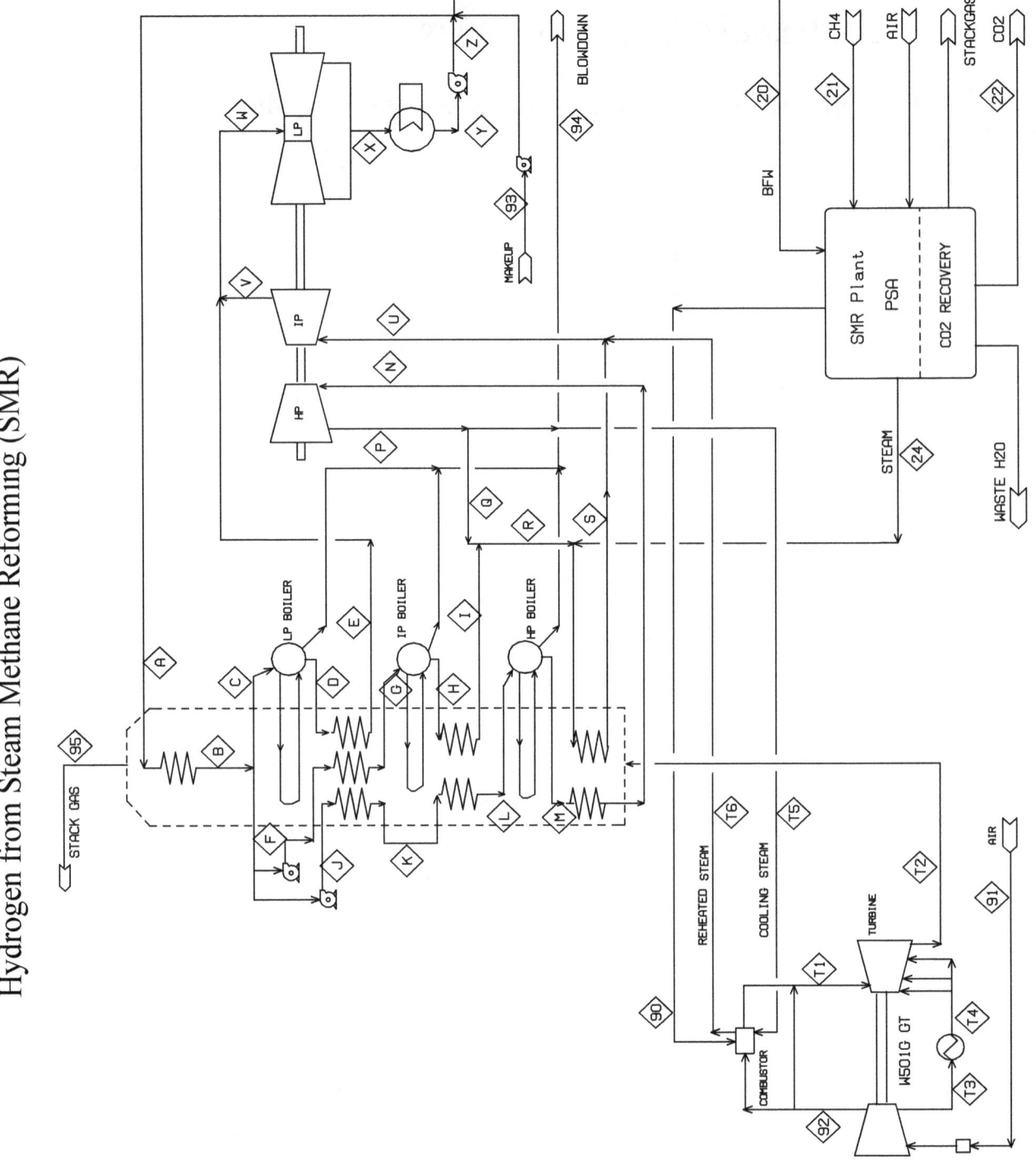

A-78

HYDROGEN TURBINE CYCLE - NATURAL GAS

	MWe	EFF:	
GAS TURBINE	269.4	LHV %	64.4
STEAM TURBINE	174.8	HHV %	54.7
MISC/AUX	14.0		
SMR	3.6		(based on CH4)
CO2 CPR	13.5		42.9
NET POWER	413.1		38.6

Stream PFD #	A	B	C	D	E	F	G	H	I	J	K	L	M	N	P
ASPEN Name ID	TOLPEC	HOTLP	TOLPEV	TOLPSH	LPTOIP	TOIPEC	TOIPEV	TOIPSH	FRIPSH	TOHPEC1	TOHPEC2	TOHPEV	TOHPSH	TOHPTUR	FRHPTUR
Temperature F	92	295	295	299.3	400	296.9	463	497.5	615	300.2	463	615	631.5	1050	759.9
Pressure psi	73.5	66.3	66.3	66.3	63	737	700	665	632	2263.8	2150.7	2043.1	1941	1800	632
Mass Flow lb/hr	667412	667412	143488	142054	142054	155514	155514	153959	153959	368410	368410	368410	364726	364726	364726
Mole Flow lbmol/hr	37047	37047	7965	7885	7885	8632	8632	8546	8546	20450	20450	20450	20245	20245	20245
Enthalpy MMBtu/hr	-4542.3	-4406.1	-947.3	-807.9	-800.3	-1026.2	-998.6	-872	-857.4	-2428.7	-2365.5	-2295.1	-2087.6	-1953.1	-1999.4

Stream PFD #	R	S	U	V	W	X	Y	Z	90	91	92	93	94	95	T1
ASPEN Name ID	TOREHT	40	TOIPTUR1	TOIPMX2	TOIPTUR2	TOCOND	TOCPMP	TOCMIX	FLH2	1	2	MAKUP	TBLOW	GTPC9	31
Temperature F	702.7	1050	1054.9	519.9	504.5	93.6	90	90.1	325	59	813.2	80	213	208.5	2583.1
Pressure psi	632	600	600	63	63	0.8	0.7	73.5	350	14.7	282.2	20	15	15	268.5
Mass Flow lb/hr	887785	887785	977785	977785	1119839	1119839	1119839	1119839	45203	4320000	3785688	29629	6674	4365208	3830896
Mole Flow lbmol/hr	49279	49279	54275	54275	62160	62160	62160	62160	21157	149707	131191	1645	370	160336	141820
Enthalpy MMBtu/hr	-4896.2	-4724.5	-5200.9	-5450.1	-6250.3	-6560.5	-7623.8	-7623.5	35.8	-180.4	551	-202	-42.2	-2201.7	547.9

Stream PFD #	T3	T4	T5	T6	20	21	22	24
ASPEN Name ID	3	12	C3	C4	TOREFORM	CH4R	CO2CAL	32
Temperature F	813.2	600	759.9	1103.2	89.8	60	123	700
Pressure psi	282.2	277	632	600	73.5	150	3000	632
Mass Flow lb/hr	527109	527109	90000	90000	482056	152843	297040	459101
Mole Flow lbmol/hr	18267	18267	4996	4996	26758	9527	6749	25484
Enthalpy MMBtu/hr	76.7	47.9	-493.4	-476.3	-3281.8	-307.4	-1168.2	-2532.7

Hydrogen Turbine - CO$_2$ Capture

Destec High Pressure (E-Gas^{TM}) / HGCU / HSD

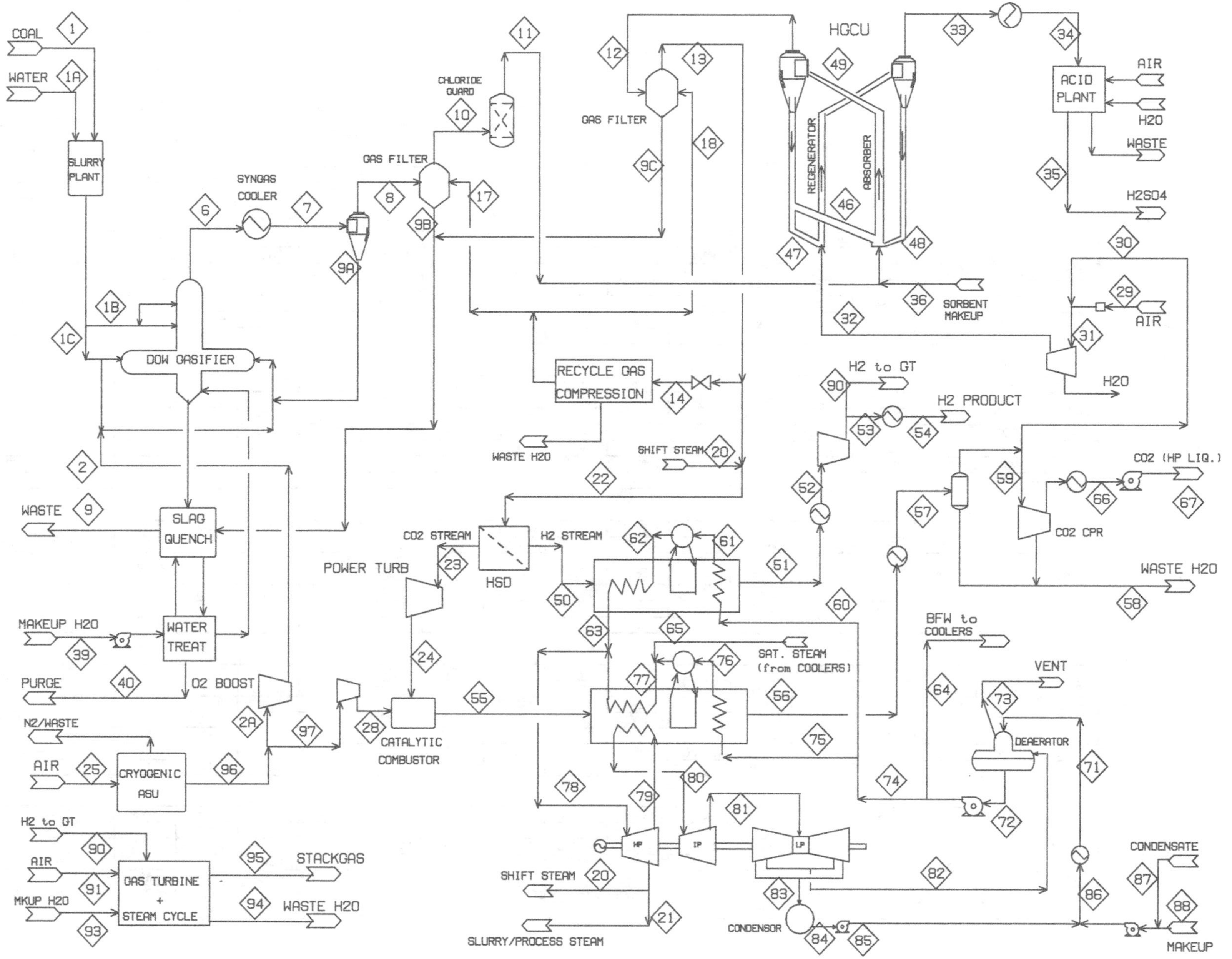

COAL

WATER

SLURRY PLANT

DOW GASIFIER

SYNGAS COOLER

GAS FILTER

CHLORIDE GUARD

GAS FILTER

RECYCLE GAS COMPRESSION

SLAG QUENCH

WATER TREAT

WASTE

MAKEUP H2O

PURGE

N2/WASTE

CRYOGENIC ASU

AIR

O2 BOOST

GAS TURBINE + STEAM CYCLE

STACKGAS

WASTE H2O

H2 to GT

AIR

MKUP H2O

CATALYTIC COMBUSTOR

POWER TURB

CO2 STREAM

HSD

H2 STREAM

SHIFT STEAM

WASTE H2O

HGCU

ABSORBER

REGENERATOR

SORBENT MAKEUP

ACID PLANT

AIR

H2O

WASTE

H2SO4

AIR

H2O

CO2 (HP LIQ.)

CO2 CPR

WASTE H2O

H2 PRODUCT

H2 to GT

VENT

BFW to COOLERS

DEAERATOR

CONDENSATE

MAKEUP

SAT. STEAM (from COOLERS)

CONDENSOR

LP

IP

HP

SHIFT STEAM

SLURRY/PROCESS STEAM

A-81

HYDROGEN TURBINE CYCLE - COAL

	MWe
GAS TURBINE	269.5
STEAM TURBINE	167.2
EXPANDER	65.0
CO2 SEQ.	-31.6
H2 CPR	-29.1

	MWe
MISC	-54.2
GROSS WORK	386.9
AUX (3%)	-11.6
NET POWER	375.3

EFFICIENCY:	
LHV %	38.0
HHV %	36.6

Stream Table (streams 1–10)

Stream PFD #	1	1A	1B	1C	2	2A	6	7	8	9A	9B	9C	9	39	40	10
ASPEN Name ID	COLIN	WAT1	COLB	COLA	GOXYG	GO2A	DRXROUT	RAWPRD	DRAWGAZ	FNES	16	19	WSTSOL	MWATG	PURGE	17
Temperature F	59	59	350	350	289.4	60	1905	1110	1019	1110	1099.4	1130.8	200	59	200	1099.4
Pressure psi	14.7	14.7	1078	1078	1150	18	1034	1024	1019	1019	14.7	14.7	14.7	15	15	1000
Mass Flow lb/hr	299868	99918	83629	316157	199814	199814	581450	581450	571323	10127	1069	557	45135	36338	21109	578388
Mole Flow lbmol/hr		5546			6237	6237								2017	1172	
Enthalpy MMBtu/hr	-938.8	-688	-186.1	-696.9	8.5	-0.8	-925.3	-1131.1	-1118.1	-13	-1.4	-2.1	-167.7	-250.2	-142.1	-1135.4

Stream Table (streams 11–29)

Stream PFD #	11	12	13	14	17	18	20	21	22	23	24	25	96	97	28	29
ASPEN Name ID	18	20	26	21	25	24C	SHFSTM	SLURSTM	TOSHF1	CO2RICH	S5	AIRASU	10	O2CAT	N862	RAIR
Temperature F	1095.4	1136.6	1130.8	1130.8	334.3	334.3	875	879.1	1014.4	1391.9	555.6	59	60	80	159.1	59
Pressure psi	985	975	965	965	1291.2	1291.2	1000	1000	964	950	20.5	14.6	18	16.5	25	14.6
Mass Flow lb/hr	577509	574248	578078	12712	8135	4387	303352	87912	868726	823336	823336	1193496	263168	63356	63356	47049
Mole Flow lbmol/hr	1174	2804	30687	675	432	233	16838	4880	46851	25690	25690	41360	8214	1977	1977	1630
Enthalpy MMBtu/hr	-1135.5	-1149.9	-1157.8	-25.5	-18.6	-10.1	-1648.9	-477.4	-2781.2	-2979	-3204.3	-49.8	-1	0	1.1	-2

Stream Table (streams 30–55)

Stream PFD #	30	31	32	33	34	35	46	47	48	49	50	51	52	53	54	55
ASPEN Name ID	30X	N830	39	5A	5	SACID	46	47	48	49	H2PRD	S10	S28	S34	H2HPPRD	CATOUT
Temperature F	60	56.7	260.9	1385.5	850	100	1135.5	1135.5	1385.5	1136.6	1391.9	300	85	324.6	190	1870.2
Pressure psi	14.8	14.6	971	955	940	16	975	975	955	954	20.5	19.6	18.5	350	346.5	19.5
Mass Flow lb/hr	50627	97676	96917	100678	100678	23173	7161983	795776	792015	8532007	45385	45385	45385		1	886692
Mole Flow lbmol/hr	1174	2804	2762	2645	2645	239	61558	6840	6606		21161	21161	21161	0	0	26022
Enthalpy MMBtu/hr	-193.3	-195.3	-186.9	-187.3	-201.9	-28.9	-24486.1	-2720.7	-2722.3	-28354.7	195	31.8	0.2	0	1	-3203.1

Stream Table (streams 56–74)

Stream PFD #	56	57	58	59	60	61	62	63	64	65	66	67	71	72	73	74
ASPEN Name ID	S11	S35	N845	TOCO2CPF	TSTMCH4	S22	S23	S25	TOSYNCOF	FRSYNCOL	14	CO2LIQ	TODEAER	TOPMPHP	DVENT	S17
Temperature F	275	80	80	80	2345.6	620	629.3	1050	221.3	635	100	123	205	217.3	217.3	221.3
Pressure psi	18.7	14.8	14.8	14.8	2345.6	2011.1	1910.5	1800	2345.6	1911	2060	3000	17.1	16.3	16.3	2345.6
Mass Flow lb/hr	886692	886691	169126	666937	120018	120018	118818	118818	231609	231609	659527	659527	604672	611688	3074	611688
Mole Flow lbmol/hr	26022	26022	9384	15465	6662	6662	6595	6595	12856	12856	15054	15054	33564	33953	171	33953
Enthalpy MMBtu/hr	-3665.6	-3895.7	-1159.9	-2542.8	-800.7	-746.7	-679.8	-636.3	-1545.2	-1322.4	-2562.7	-2558.4	-4047.1	-4086.5	-17.6	-4080.9

Stream Table (streams 75–91)

Stream PFD #	75	76	77	78	79	80	81	82	83	84	85	86	87	88	90	91
ASPEN Name ID	TSTMCO2	TOBLR	S19	S20	HPTUREX	IPTURIN	IPTUREX	LPDEAER	VLPEX	CNDOUT	TOMIX	TOCNDQ	SLURCND	MKUP	90	1
Temperature F	221.3	620	631.9	1050	709.8	1050	555.8	355	92.3	91	91	98.3	180	80	324.6	59
Pressure psi	2345.6	2011.1	1910.5	1800	518	492.1	63	17.1	0.8	0.7	20	20	20	20	350	14.7
Mass Flow lb/hr	260062	260062	489070	607887	216623	216623	216623	10090	206533	206533	206533	604672	87912	310227	45384	4320000
Mole Flow lbmol/hr	14435	14435	27147	33742	12024	12024	12024	560	11464	11464	11464	33564	4880	17220	21161	149707
Enthalpy MMBtu/hr	-1735	-1618	-2795.5	-3255.3	-1192.4	-1152.2	-1203.6	-57	-1207.5	-1405.9	-1405.8	-4111.5	-590.6	-2115.1	-180.4	

Stream Table (streams 93–97)

Stream PFD #	93	94	95	96	97
ASPEN Name ID	MAKUP	TBLOW	GTPC9	O2CAT	O2CAT
Temperature F	80	213	208.5	60	80
Pressure psi	20	15	15	18	16.5
Mass Flow lb/hr	7046	7046	4365389	263168	63356
Mole Flow lbmol/hr	391	391	160341	8214	1977
Enthalpy MMBtu/hr	-48	-44.4	-2201.9	-1	0

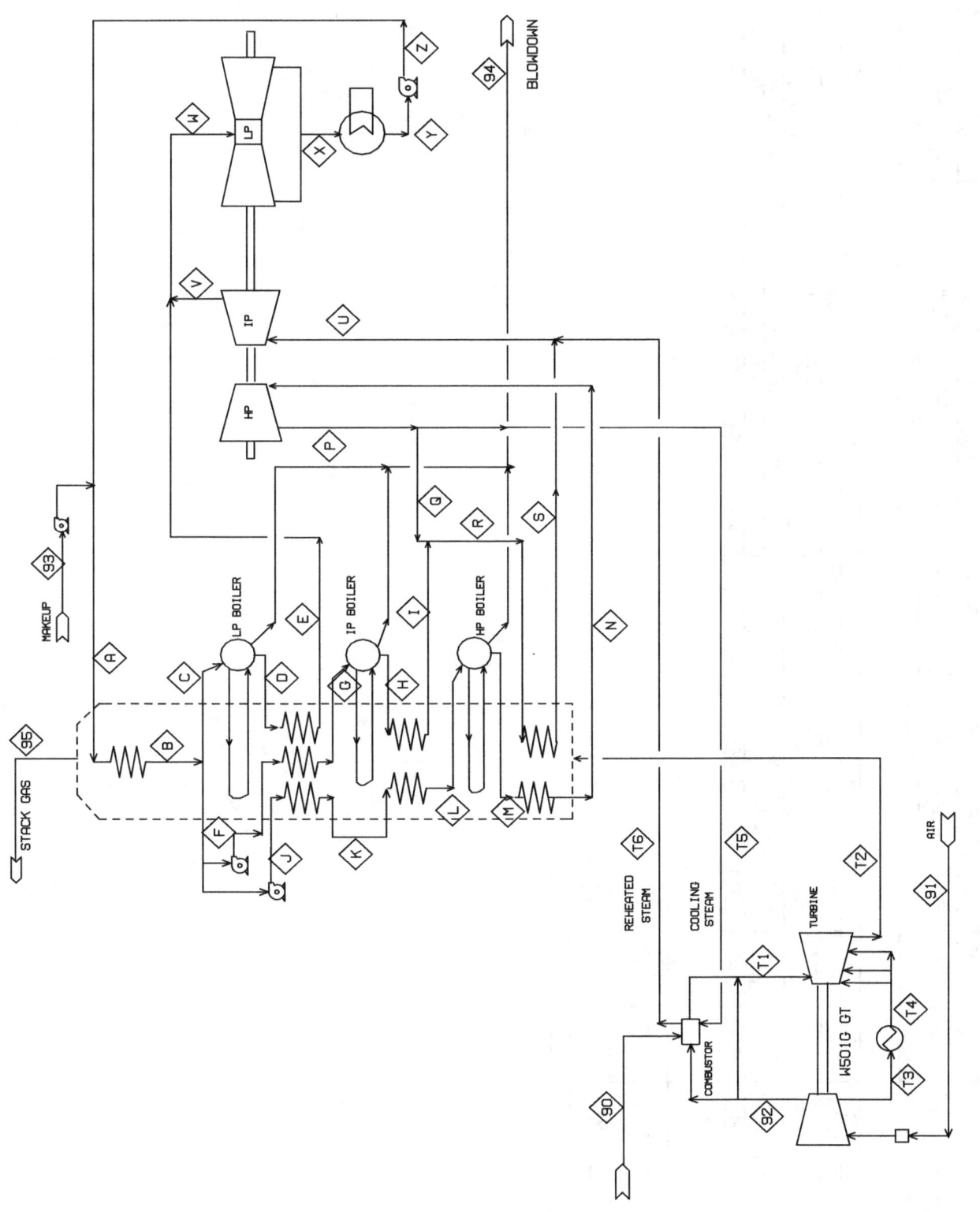

A-83

HYDROGEN TURBINE CYCLE - COAL
STEAM CYCLE

Stream PFD #	A	B	C	D	E	F	G	H	I	J	K	L	M	N	P	Q
ASPEN Name ID	TOLPEC	HOTLP	TOLPEV	TOLPSH	LPTOIP	TOIPEC	TOIPEV	TOIPSH	FRIPSH	TOHPEC1	TOHPEC2	TOHPEV	TOHPSH	TOHPTUR	FRHPTUR	TMXIP
Temperature F	92	295	295	299.3	400	296.4	463	472.8	615	300	463	615	631.5	1050	712	712
Pressure psi	73.5	66.3	66.3	66.3	63	585.7	556.4	528.6	518	2263.8	2150.7	2043.1	1941	1800	518	518
Mass Flow lb/hr	703403	703403	89167	88276	88276	170988	170988	169278	169278	443247	443247	443247	438814	438814	438814	358814
Mole Flow lbmol/hr	39044	39044	4949	4900	4900	9491	9491	9396	9396	24604	24604	24604	24358	24358	24358	19917
Enthalpy MMBtu/hr	-4787.2	-4643.7	-588.7	-502	-497.3	-1128.4	-1098	-958.5	-941.1	-2922.1	-2846	-2761.3	-2511.7	-2349.9	-2414.8	-1974.6

Stream PFD #	R	S	U	V	W	X	Y	Z	90	91	92	93	94	95
ASPEN Name ID	TOREHT	40	TOIPMX2	TOIPTUR1	TOIPTUR2	TOCOND	TOCPMP	TOCMIX	90	1	2	MAKUP	TBLOW	GTPC9
Temperature F	680.1	1050	561	1057	540.4	93.6	90	90.1	324.6	59	813.2	80	213	208.5
Pressure psi	518	492	63	492	63	0.8	0.7	73.5	350	14.7	282.2	20	15	15
Mass Flow lb/hr	528093	528093	608093	608093	696369	696369	696369	696369	45384	4320000	3785688	7034	7034	4365389
Mole Flow lbmol/hr	29313	29313	33754	33754	38654	38654	38654	38654	21161	149707	131191	390	390	160341
Enthalpy MMBtu/hr	-2915.7	-2808.8	-3377.1	-3232	-3874.4	-4072.2	-4740.9	-4740.7	35.6	-180.4	551	-48	-44.3	-2201.9

Stream PFD #	T3	T4	T5	T6
ASPEN Name ID	3	12	C3	C4
Temperature F	813.2	600	712	1103.2
Pressure psi	282.2	277	518	492
Mass Flow lb/hr	527109	527109	80000	80000
Mole Flow lbmol/hr	18267	18267	4441	4441
Enthalpy MMBtu/hr	76.7	47.9	-440.2	-423.2

Hybrid Cycles (Turbine / SOFC)

Natural Gas Hybrid Turbine / SOFC

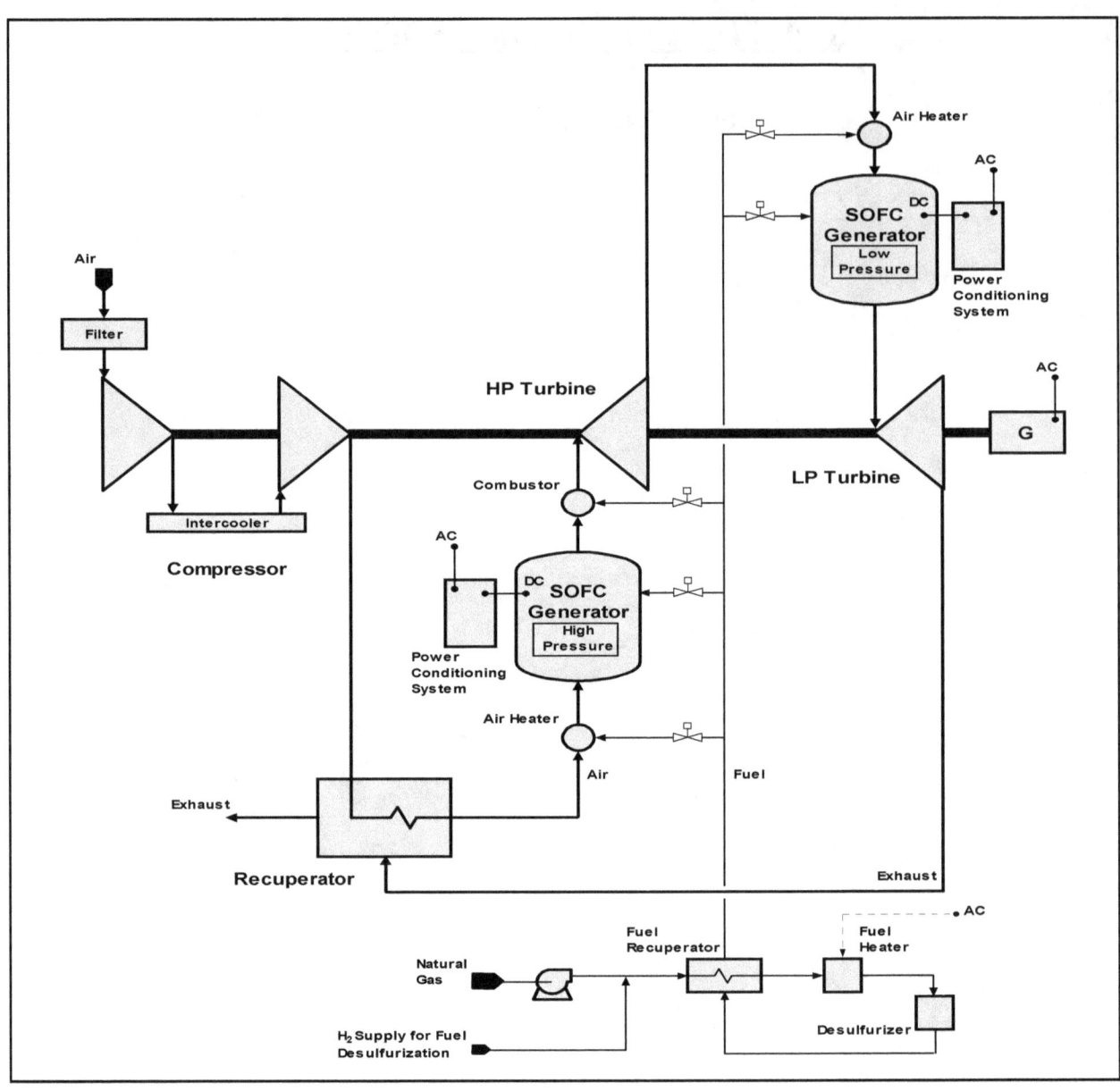

Natural Gas Hybrid M&E

See report:

"Pressurized Solid Oxide Fuel Cycle/Gas Turbine Power System" by Siemens Westinghouse / Rolls-Royce Allison for the DOE. (DE-AC26-98FT40355 , February 2000).

Hybrid Cycles (Turbine / SOFC)

Destec (E-Gas^{TM}) / HGCU / "G" GT / No CO_2 Capture

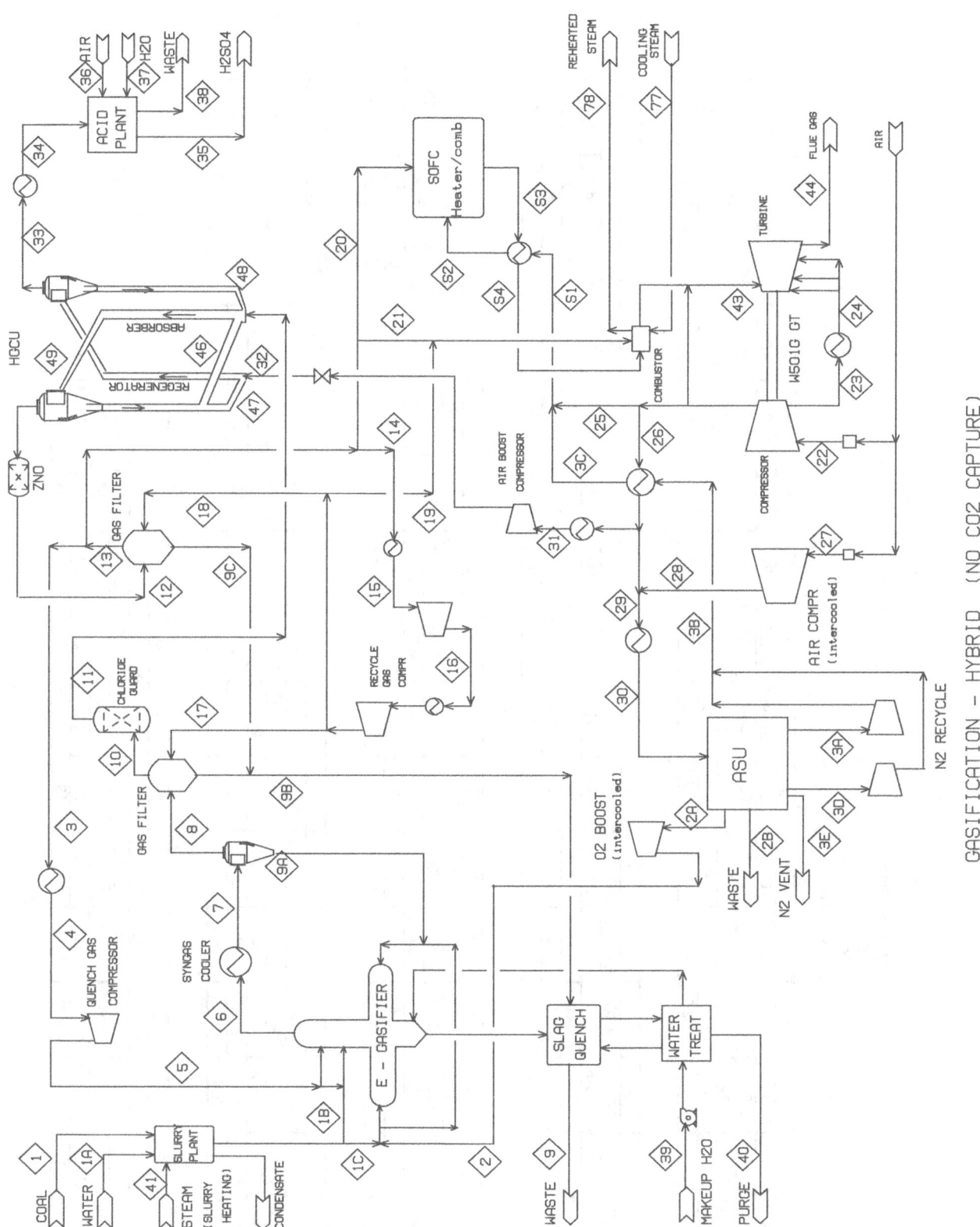

GASIFICATION - HYBRID (NO CO2 CAPTURE)

A-89

4/29/2002

E-GAS (DESTEC) GASIFICATION - HYBRID POWER SYSTEM
(GAS TURBINE / HGCU/SOFC/STEAM CYCLE)
(NO CO2 SEQUESTRATION)
(58% syngas to SOFC)

POWER kW			
GAS TURBINE	-276.1	LHV EFFICIENCY	56.4 %
SOFC	-221.4	HHV EFFICIENCY	54.4 %
STEAM TURBINE	-207.7		
MISC	41.6		
AUX	19.9		
NET POWER	-643.6		

Stream PFD ID	1	1A	1B	1C	2A	2B	2	3A	3B	3C	3D	3E	3	4
ASPEN ID	COLIN	WAT1	COALB	COLA	GO2A	ASUWST	GOXYG	N2RCY	317	HOTN2	3D	N2OUT	RECYGAS	GRCYCX
Mass Flow lb/hr	345386	115085	96323	364148	266867	3949	266867	285936	340916	340916	54980	508331	235012	235012
Temperature F	59	59	350	350	60	59	204.7	62	183.7	700	60	62	1053.2	300
Pressure psi	14.7	14.7	465	465	92	14.7	472	91	300	294	265	91	346	336
H MMBtu/hr	-1081.3	-792.5	-206	-772.9	-1.2	-27.2	6.7	-3.1	6.7	51.6	-0.4	-5.5	-502.3	-572.9

Stream PFD ID	5	6	7	8	9	9A	9B	9C	12	11	10	41	40	39
ASPEN ID	GRCC	DRXROUT	RAWPRD	DRAWGAZ	WSTSOL	FNES	16	19	GFLT1	18	17	STOPRE	PURGE	MWATG
Mass Flow lb/hr	235016	941443	941443	929779	51410	11664	1231	65	934471	938815	939828	70699	24314	41853
Temperature F	359.3	1900	1004	1004	200	1004	997	1053.2	1057	994	997	863.9	200	59
Pressure psi	425	412	403.8	394.5	14.7	394.5	14.7	14.7	356	366	394.5	150	15	15
H MMBtu/hr	-567.7	-1622.5	-1983.7	-1968.4	-190.6	-15.3	-1.6	-0.1	-1995.6	-1993.9	-1993.8	-382.2	-163.7	-288.2

Stream PFD ID	13	14	15	16	17	18	19	20	21	S1	S2	S3	S4	22
ASPEN ID	26	21	22	23	25C	24	27	TOFCELL	TOGT	C1	CATHIN	FLCEXIT	C3	AIR1
Mass Flow lb/hr	940047	18801	18801	18801	11281	5640	1880	399106	289008	3331258	3331258	3730363	3730363	4467600
Temperature F	1053.2	1053.2	300	436.1	409.4	409.4	409.4	1051.5	1051.5	801.6	1175	2070.4	1780.9	59
Pressure psi	346	346	336	565.6	900	900	900	345	345	282.2	273.8	260.1	252.3	14.6
H MMBtu/hr	-2009	-40.2	-45.8	-44.9	-27.1	-13.5	-4.5	-853.2	-617.9	486.9	816.8	-799.6	-1129.6	-186.5

Stream PFD ID	23	24	25	26	27	28	29	30	31	32	33	34	35	36
ASPEN ID	TOCHILL	COLAIR	AIR7	TOOXYG	ASU1	ASU6	AIRSUP	O2INX	REGENAIR	REGENAIR	5A	5	SACID	ACAIR
Mass Flow lb/hr	545119	545119	2990342	642103	560311	557904	1118215	1118215	81792	81792	86116	86116	26391	18728
Temperature F	813.3	600	813.3	813.3	59	203.8	373.3	190	120	167	1443.2	850	100	59
Pressure psi	282.2	276.6	282.2	282.2	14.6	278	278	275	275.2	371	361	344	16	14.7
H MMBtu/hr	79.4	49.5	435.3	93.5	-23.4	9.3	51.6	0.9	-2.6	-1.5	-8.5	-21.7	-33.2	-0.8

Stream PFD ID	37	38	43	44	46	47	48	49	68	71
ASPEN ID	ACWAT	WGAS	POC3	GTPOC	46	47	48	49	TOGAS	FRGAS
Mass Flow lb/hr	4730	83178	4295468	4840586	5997540	666393	662070	7598425	602181	602181
Temperature F	59	100	2582.8	1185.2	1055	1055	1443.2	1059	2116.9	1911
Pressure psi	14.7	16	15.2	15.2	356	356	361	361	420	635
H MMBtu/hr	-32.6	-2.6	-1741.3	-3484	-20594.1	-2288.2	-2283.2	-24871.2	-3894.7	-3438.2

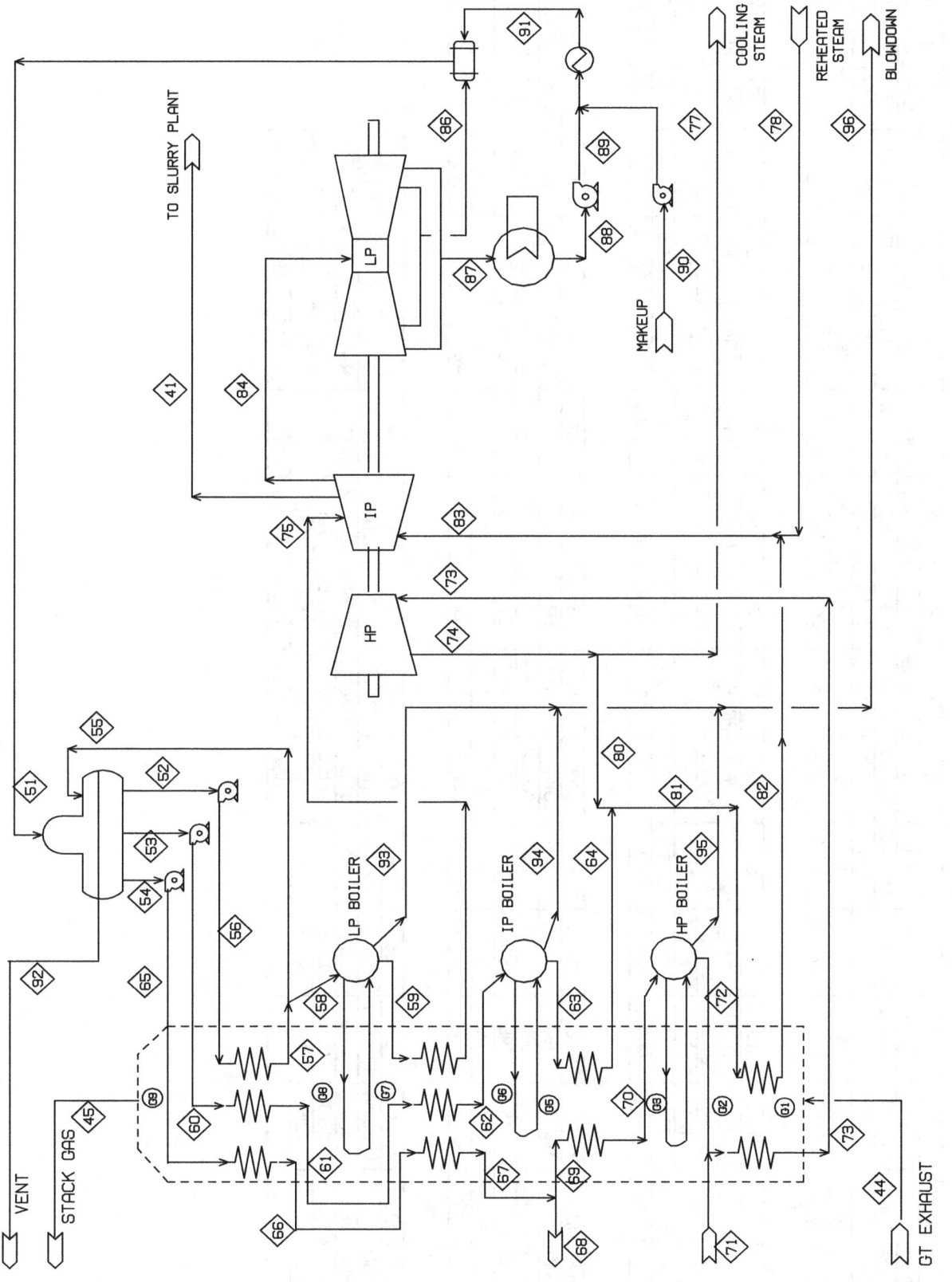

A-91

E-GAS (DESTEC) GASIFICATION - HYBRID POWER SYSTEM
(STEAM CYCLE)
(NO CO2 SEQUESTRATION)
(58% syngas to SOFC)

Stream PFD ID	41	44	45	51	52	53	54	55	56	57	58	59	60	61
ASPEN ID	STOPRE	GTPOC	GTPC9	TODEAER	TOLP	TOIPPMP	TOHPPMP	RDEAER	TOLPEC	FRLPEC	TOLPEV	LPTOIP	TOIPEC1	TOIPEC2
Mass Flow lb/hr	70699	4840587	4840587	1103427	304518	263556	819617	291237	304518	304518	13280	13148	263556	263556
Temperature F	863.9	1185.2	256.7	205	217.3	217.3	217.3	286	217.4	286	286	305.3	218.1	286
Pressure psi	150	15.2	15	17	16.3	16.3	16.3	76.3	80.3	76.3	76.3	72.5	410.6	390
H MMBtu/hr	-382.2	-3484	-4707.8	-7385.3	-2034.4	-1760.7	-5475.6	-1925.3	-2034.3	-2013.1	-87.8	-74.7	-1760.3	-1742.2

Stream PFD ID	62	63	64	65	66	67	68	69	70	71	72	73	74	75
ASPEN ID	TOIPEV	TOIPSH	FRIPSH	TOHPEC1	TOHPEC2	FRHPEC2	TOGAS	TOHPEC3	TOHPEV	FRGAS	TOHPSH	TOHPTUR	FRHPTUR	713
Mass Flow lb/hr	263556	260921	260921	819617	819617	819617	602181	217436	217436	602181	215262	817443	817443	13148
Temperature F	420	432.3	620	221.1	286	420	420	420	620	635	629.3	1099.3	645	420
Pressure psi	370.5	352	350	2345.6	2228.3	2116.9	2116.9	2116.9	2011.1	1911	1910.5	1800	350	69.5
H MMBtu/hr	-1705.1	-1477.5	-1446.5	-5468.3	-5415	-5301	-3894.7	-1406.3	-1352.8	-3438.2	-1231.6	-4353.1	-4520.5	-73.9

Stream PFD ID	77	78	80	81	82	83	84	86	87	88	89	90	91	92
ASPEN ID	TOSTAT	FRGT	FRHPS	TOREHT	TOIPMIX	TOIPTUR1	TOLPTUR1	314	TOCOND	TOCPMP	TOFWH	MAKUP	FRFWHTR	DEBLOW
Mass Flow lb/hr	70000	70000	747443	1008364	1008364	1078364	1020812	39378	981434	981434	981434	82615	1064049	6973
Temperature F	645	1095.6	645	638.5	1100	1099.7	515.7	379.7	88.8	87.9	87.9	80	165.9	217.3
Pressure psi	350	342	350	350	342	342	35	17	0.7	0.7	40	14.7	17	16.3
H MMBtu/hr	-387.1	-370.3	-4133.4	-5579.9	-5332.2	-5702.5	-5689.5	-222	-5724.3	-6683.7	-6683.5	-563.3	-7163.4	-39.8

Stream PFD ID	93	94	95	96	G1	G2	G3	G5	G6	G7	G8	G9
ASPEN ID	LPBLOW	IPBLOW	HPBLOW	TBLOW	GTPC1	GTPC2	GTPC3	GTPC5	GTPC6	GTPC7	GTPC8	GTPC9
Mass Flow lb/hr	133	2636	2174	4943	4840586	4840586	4840586	4840586	4840586	4840586	4840586	4840587
Temperature F	305.3	432.3	629.3	213	1185.2	772.1	690.5	625.7	461.7	341	331.3	256.7
Pressure psi	72.5	352	1910.5	15	15.2	15.2	15.2	15.2	15.2	15.2	15.2	15
H MMBtu/hr	-0.9	-17	-13.5	-31.4	-3484	-4048.5	-4156.1	-4240.6	-4451.2	-4603.1	-4615.3	-4707.8

Hybrid Cycles (Turbine / SOFC)

Destec High Pressure (E-Gas^{TM}) / HGCU / "G" GT / CO_2 Capture

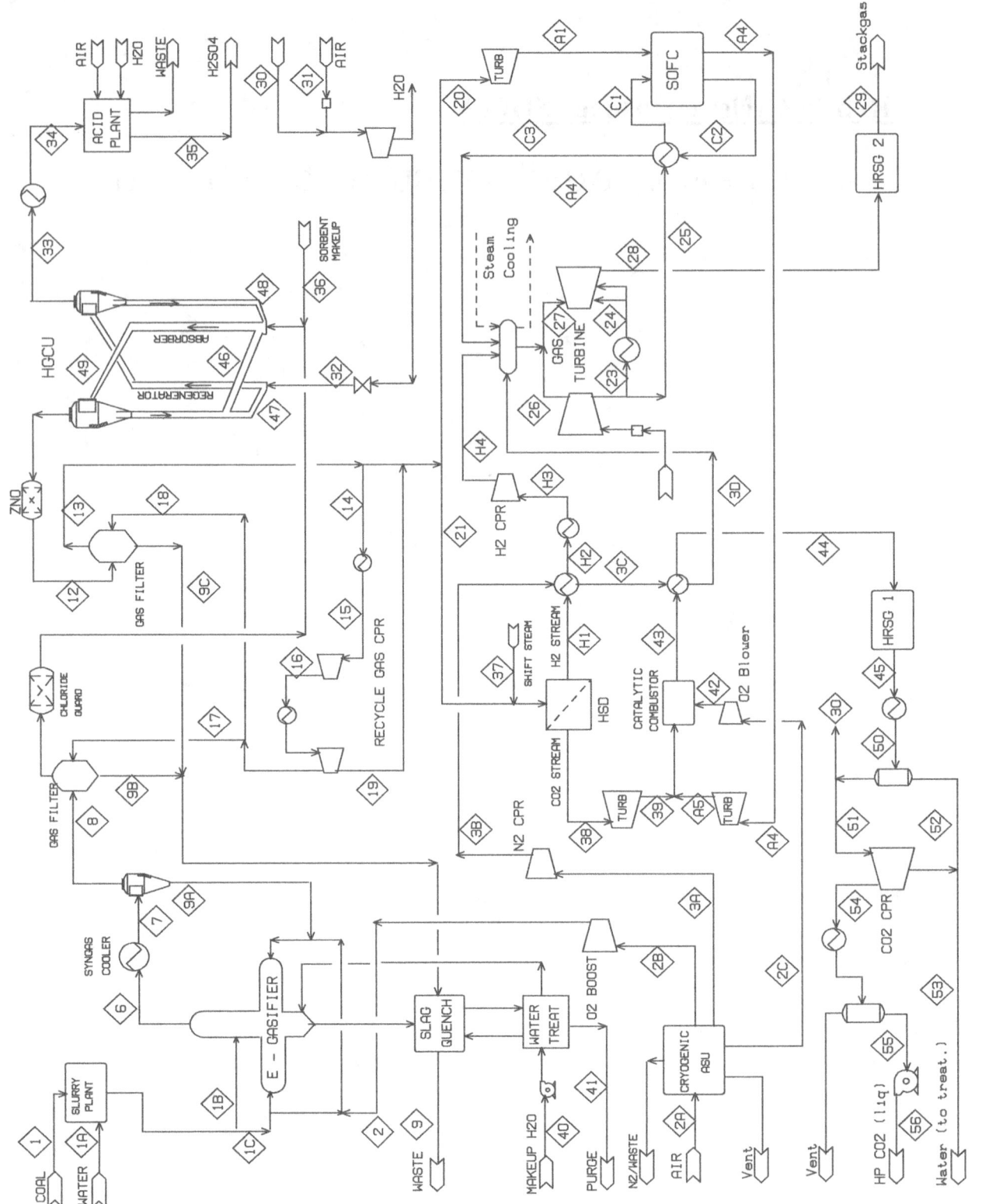

A-94

4/30/2002

E-GAS (DESTEC) GASIFICATION - HYBRID POWER SYSTEM
(GAS TURBINE / HGCU//SOFC/STEAM CYCLE)
(CO2 SEQUESTRATION)

POWER kW

GAS TURBINE	-272.5	LHV EFFICIENCY	49.7 %
SOFC	-324.1	HHV EFFICIENCY	47.9 %
STEAM TURBINE	-226.1		
MISC (generated)	-121.2		
MISC (required)	166.1		
AUX	23.3		
NET POWER	-754.6		

Stream PFD ID	1	1A	1B	1C	2A	2b	2	2C	3A	3B	3C	3D	6	7	8
ASPEN ID	COLIN	WAT1	COLB	COLA	AIRASU	GO2A	GOXYG	O2LP	3A	35	68	HOTN2	DRXROUT	RAWPRD	DRAWGAZ
Mass Flow lb/hr	460812	153546	128514	485844	1836831	307420	307420	97608	460000	458841	458841	458841	893884	893884	878321
Temperature F	59	59	350	350	59	60	289.4	60	60	287.1	1300	1900	1904.7	1110	1110
Pressure psi	14.7	14.7	1078	1078	14.6	18	1150	16.5	18	260	255	250	1034	1024	1019
H MMBtu/hr	-1442.7	-1057.3	-286	-1070.9	-76.7	-1.2	13	-0.4	-23.4	8.8	131	209.2	-1421.9	-1738.1	-1718.1

Stream PFD ID	9B	9C	9	40	41	10	11	12	13	14	15	16	17	18	19
ASPEN ID	16	19	WSTSOL	MWATG	PURGE	17	18	TOGFLT	26	21	22	23	25C	24	27
Mass Flow lb/hr	1643	856	69360	55841	32439	888029	886677	881653	886918	17738	17738	17676	11351	6121	17
Temperature F	1100.4	1134.6	200	59	200	1100.4	1096.4	1139.9	1134.6	1134.6	300	344.9	334.3	334.3	334.3
Pressure psi	14.7	14.7	200	59	15	1000	985	975	965	965	955	1146	1291.2	1291.2	1291.2
H MMBtu/hr	-2.1	-3.3	-257.7	-384.5	-218.4	-1742.1	-1742.2	-1763.4	-1774.2	-35.5	-41.8	-41	-26	-14	0

Stream PFD ID	21	A1	A4	A5	C1	C2	C3	22	23	24	25	26	27	28	29
ASPEN ID	TOSHFT	A1	A4	A5	C1	C2	C3	AIR1	TOCHILL	COLAIR	2AX	2B	TOHPGT	GTPCX	34X
Mass Flow lb/hr	365061	504131	799622	799622	3638756	3343264	3343264	4467600	545119	545119	3638756	276276	4107854	4660422	4660422
Temperature F	1134.6	777.8	1831.9	1176.5	1075	1832	1567.6	59	813.2	600	813.2	813.2	2582.6	1105.6	208.5
Pressure psi	964	282.3	252.3	25	273.8	260.1	252.3	14.6	282.2	277	282.2	282.2	242.2	15	14.7
H MMBtu/hr	-730.3	-1085.6	-2837.1	-3034.3	780.7	1415	1163.9	-186.6	79.3	49.5	529.6	40.2	1403.8	-325.3	-1447.3

Stream PFD ID	31	32	33	34	35	36	37	38	39	H1	H2	H3	H4	42	43
ASPEN ID	RAIR	39	5A	5	SACID	ZNMKUP	SHFSTM	CO2RICH	CO2CMB	H2PRD	S10	S28	H2GT	O2CAT	CATOUT
Mass Flow lb/hr	72295	148931	154711	154711	35607	770	195946	531518	531518	29485	29485	29485	29485	97608	1428748
Temperature F	59	260.9	1391.8	850	100	100	875	1393.5	556.2	1393.5	127.7	85	312	136.4	2142.9
Pressure psi	14.6	971	955	940	16	985	1000	950	20.5	20.5	19.6	18.5	300	25	19.5
H MMBtu/hr	-3	-287.3	-287.6	-310.3	-44.3	-3.4	-1065.1	-1923.4	-2069	126.3	4.1	0.1	21.8	1.3	-5102

Stream PFD ID	45	46	47	48	49	50	51	52	53	54	55	56
ASPEN ID	S11	46	47	48	49	S35	TOCO2CPF	FLSH2O	WSTH2O	CO2PROD	29	CO2LIQ
Mass Flow lb/hr	1428748	11005772	1222864	1217084	13110303	1428746	1023486	327456	338819	1012124	1012124	1012124
Temperature F	275	1139.2	1139.2	1391.8	1139.9	80	80	80	80.7	268.3	100	122.9
Pressure psi	18.7	975	975	955	954	14.8	14.8	14.8	14.8	2100	2060	3000
H MMBtu/hr	-6017.8	-37618.5	-4179.8	-4181.5	-43559.8	-6451.1	-3907.7	-2246.4	-2323.9	-3837.8	-3937.7	-3931.3

A-95

STEAM CYCLE
(CO2 SEQUESTRATION)

A-96

E-GAS (DESTEC) GASIFICATION - HYBRID POWER SYSTEM
(STEAM CYCLE)
(CO2 SEQUESTRATION)

Stream PFD ID	57	58	59	60	61	62	63	64	65	66	67	68	69	70
ASPEN ID	TOLPEC	HOTLP	TOLPEV	TOLPSH	LPTOIP	TOIPEC	TOIPEV	TOIPSH	FRIPSH	6	TOREHT	TOIPTUR1	TOHPEC1	TOHPEC2
Mass Flow lb/hr	734450	734450	90447	89542	89542	176219	176219	174457	174457	383106	557564	557564	467784	467784
Temperature F	90	295	295	299.3	400	296.4	463	472.8	615	712	680.8	1050	299.9	463
Pressure psi	73.5	66.3	66.3	66.3	63	585.7	556.4	528.6	518	518	518	492	2263.9	2150.7
H MMBtu/hr	-5000	-4848.6	-597.1	-509.2	-504.4	-1162.9	-1131.6	-987.8	-969.9	-2108.3	-3078.2	-2965.6	-3083.9	-3003.5

Stream PFD ID	71	72	73	74	75	76	77	78	79	80	81	82	83	84
ASPEN ID	TOHPEV	TOHPSH	TOHPTUR	3	MAKUP	LPTODEA	TOMIX	SLURCND	MKUP	TODEAER	TOPMPHP	TSTMCO2	TOSYNCOI	TOBLR
Mass Flow lb/hr	467784	463106	463106	727106	7345	13900	480230	135216	204856	820301	830031	473875	356156	473875
Temperature F	615	631.5	1050	90	80	322	91	180	80	205	217.3	221	221	620
Pressure psi	2043.1	1941	1800	90	20	17.5	20	20	20	17.1	16.3	2345.6	2345.6	2011.1
H MMBtu/hr	-2914.1	-2650.7	-2480	-4950.1	-50.1	-78.7	-3268.9	-908.4	-1396.7	-5490.3	-5545.2	-3161.6	-2376.2	-2948.2

Stream PFD ID	85	86	87	88	89	90	91	92	93	94	95	96	97	98
ASPEN ID	FRSYNCOI	S19	HPTURIN	HPTUREX	IPTURIN	COLSTM	REHSTM	SLURSTM	LPBLOW	IPBLOW	HPBLOW	TBLOW	DVENT	BLDWN
Mass Flow lb/hr	356156	825292	825292	494130	494130	80000	80000	135216	904	1762	4678	7345	4171	4739
Temperature F	635	631.7	1050	709.8	1050	712	1050	879.1	299.3	472.8	631.5	213	217.3	629.3
Pressure psi	1911	1910.5	1800	518	492.1	518	492	1000	66.3	528.6	1941	15	16.3	1910.5
H MMBtu/hr	-2033.5	-4717.7	-4419.5	-2719.8	-2628.2	-440.2	-425.5	-734.2	-6	-11.3	-29	-46.3	-23.8	-29.4

Stream PFD ID								3	S1	S2	S3	S4	S5	S6
ASPEN ID	TOIPMX2	IPTUREX	TOCOND	VLPEX	TOCPMP	CNDOUT	TOMIX	3						
Mass Flow lb/hr	637564	494130	727106	480230	727106	727106	480230	727106	1288398	957236	1131694	1131694	1221236	1207336
Temperature F	555.8	555.8	93.6	92.3	90	90	91	90	1050	712	1050	555.8	544.3766	92.3
Pressure psi	63	63	0.8	0.8	0.7	0.7	20	20	1800	518	492	63	63	0.8
H MMBtu/hr	-3542.4	-2745.5	-4253.6	-2807.6	-4950.1	-4950.1	-3268.9	-4950.1	-6899.5	-5268.3	-6019.3	-6287.9	-6792.3	-7061.2

Stream PFD ID	S7	S8
ASPEN ID		
Mass Flow lb/hr	1207336	1207336
Temperature F	90	90
Pressure psi	0.7	20
H MMBtu/hr	-8219	-8219

Hybrid Cycles (Turbine / SOFC)

Destec (E-Gas^{TM}) / OTM / CGCU / "G" GT / No CO_2 Capture

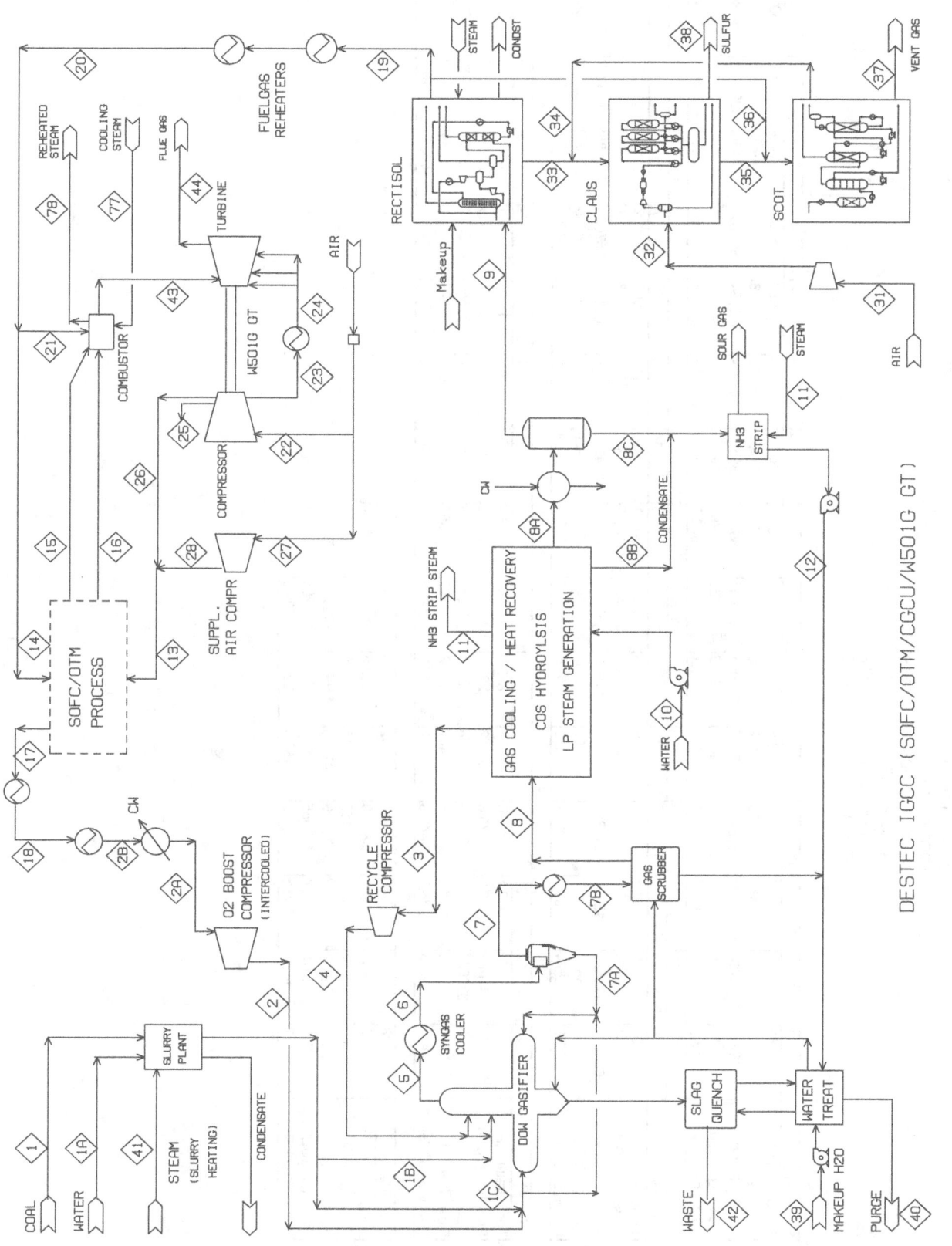

DESTEC IGCC (SOFC/OTM/CGCU/W501G GT)

A-99

E-GAS (DESTEC) GASIFICATION - HYBRID POWER SYSTEM
(NO CO2 SEQUESTRATION)

SUMMARY:	MWe	EFFICIENCY:	%
GAS TURBINE	272.7	LHV	57.02
STEAM TURBINE	189.8	HHV	54.99
SOFC POWER	254.4		
MISCELLANEOUS	-20.9		
GROSS POWER	696		
AUXILIARY (3%)	-20.9		
NET POWER	675.2		

STREAM	1	1A	1B	1C	2	2A	2B	3	4	5	6	7	7A	7B	8
Temperature (F)	59	59	350	350	223.8	80	140	303.7	332.9	1900	650	649.9	649.9	415	303.3
Pressure (PSIA)	14.7	14.7	465	465	472	10	11	378	425	412	403.8	394.5	394.5	390	380
Flow (LB/HR)	359277	119714	100197	378794	257095	257095	257095	189561	189560	903894	903894	891761	12134	891761	947803
Flow (LBMOL/HR)		6645			8035	8035	8035	9795	9795						48977
H (MM BTU/HR)	-1124.8	-824.3	-211.5	-794.3	7.7	0.1	3.5	-502.8	-500.7	-1598.2	-2082.1	-2065	-17.1	-2148.6	-2514.1

STREAM	8A	8B	8C	9	10	11	12	13	14	15	16	17	18	19	20
Temperature (F)	190	231.9	101.8	103	59	280	212.4	812.1	790	790	1660	1661.2	588	116	790
Pressure (PSIA)	354	354	20	349	14.7	37	470	282.2	330	330	276	17.5	12.5	340	330
Flow (LB/HR)	634804	123439	15499	619305	62183	62183	137792	3818369	344226	0	3905498	257095	257095	584357	584357
Flow (LBMOL/HR)	32331	6850	830	31501	3452	3452	7649	132322	17950	0	134096	8035	8035	30472	30472
H (MM BTU/HR)	-1335.9	-826.4	-99.7	-1272	-428.2	-353.6	-925.7	555.2	-606.4	0	-1034	100.1	30	-1182.6	-1029.4

STREAM	21	22	23	24	25	26	27	28	31	32	33	34	35	36	37
Temperature (F)	790	59	812.1	600	812.1	812.1	59	812.4	59	161	156.2	70	419.6	116	70
Pressure (PSIA)	330	14.6	282.2	276.6	282.2	282.2	14.6	282.2	14.7	25	18.5	17.5	26.7	340	17.5
Flow (LB/HR)	240131	4320000	527109	527109	13478	3779413	38956	38956	19759	19759	32640	2589	46154	6712	50276
Flow (LBMOL/HR)	12522	149706	18266	18266	467	130972	1350	1350	685	685	923	61	1530	350	1816
H (MM BTU/HR)	-423	-180.5	76.6	47.9	2	549.6	-1.6	5.7	-0.8	-0.3	-97.7	-9	-130.6	-13.6	-149.2

STREAM	38	39	40	41	42	43	44	45
Temperature (F)	285	59	200	819.9	200	2583.5	1135.2	259.7
Pressure (PSIA)	14.7	14.7	15	150	15	268.5	15.2	15
Flow (LB/HR)	8834	52508	120578	73226	47563	4145629	4672738	4672738
Flow (LBMOL/HR)	276	2915	6684	4065		140938	159204	159204
H (MM BTU/HR)	-0.9	-361.5	-809.3	-397.8	-158.5	-1505.6	-3209.9	-4311.7

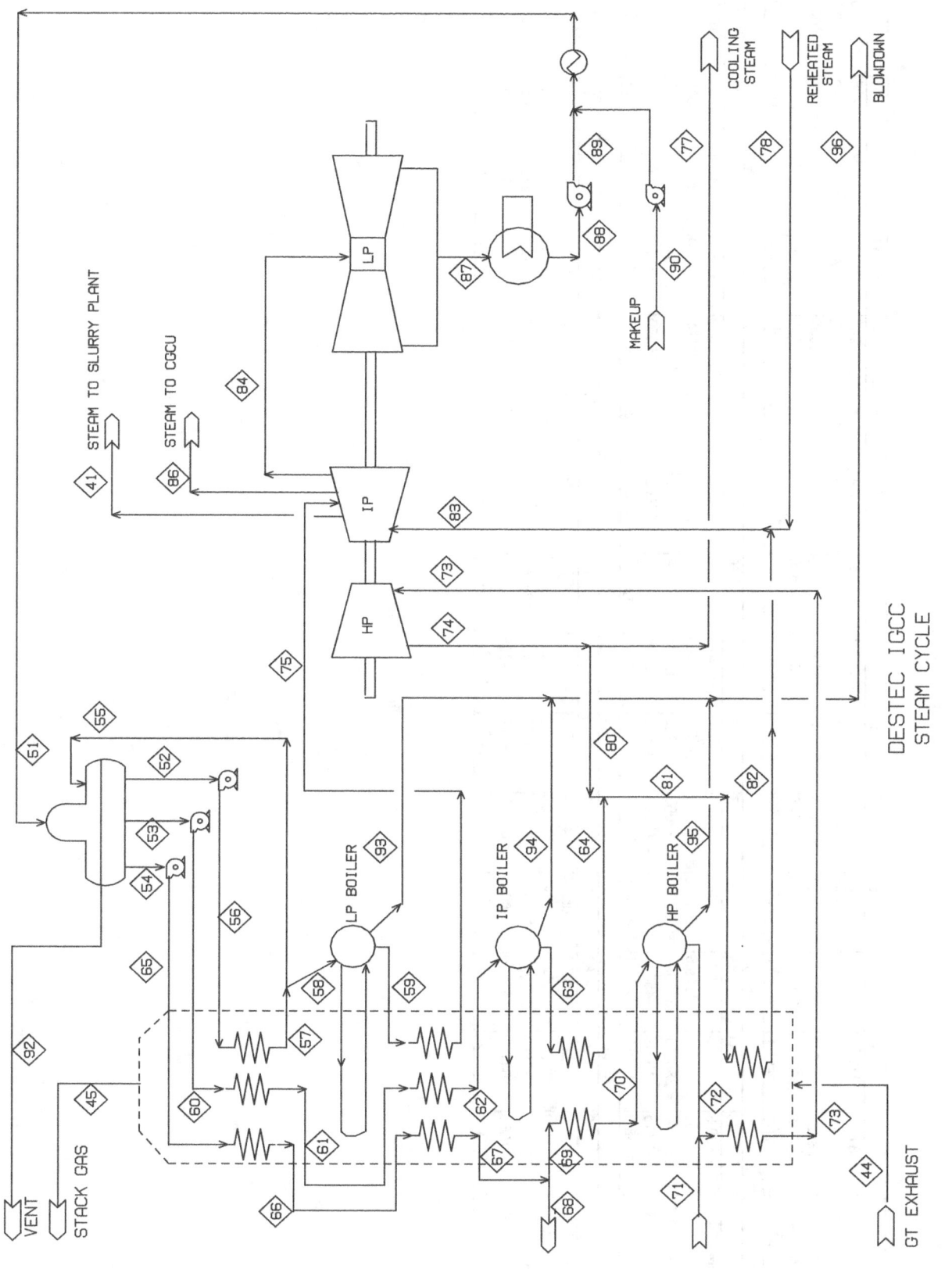

DESTEC IGCC
STEAM CYCLE

A-101

E-GAS (DESTEC) GASIFICATION - HYBRID POWER SYST
(NO CO2 SEQUESTRATION)

STEAM CYCLE PROCESS STREAMS

STREAM	41	44	45	51	52	53	54	55	56	57	58	59	60	61	62
Temperature (F)	819.9	1135.2	259.7	205	217.3	217.3	217.3	286	217.4	286	286	305.3	218.1	286	420
Pressure (PSIA)	150	15.2	15	17	16.3	16.3	16.3	76.3	80.3	76.3	76.3	72.5	410.6	390	370.5
Flow (LB/HR)	73226	4672738	4672738	1080557	298115	238664	822063	285113	298115	298115	13001	12871	238664	238664	238664
Flow (LBMOL/HR)	4065	159204	159204	59979	16548	13248	45631	15826	16548	16548	722	714	13248	13248	13248
H (MM BTU/HR)	-397.8	-3209.9	-4311.7	-7236.7	-1992.9	-1595.4	-5495.4	-1886	-1992.8	-1972	-86	-73.2	-1595	-1578.6	-1545

STREAM	63	64	65	66	67	68	69	70	71	72	73	74	75	77	78
Temperature (F)	432.3	620	221	286	420	420	420	620	635	629.3	1050	606.4	420	606.4	1056
Pressure (PSIA)	352	350	2345.6	2228.3	2116.9	2116.9	2116.9	2011.1	1910.5	1910.5	1800	350	69.5	350	342
Flow (LB/HR)	236277	236277	822063	822063	822063	676759	145304	145304	676759	143851	820610	820610	12871	70000	70000
Flow (LBMOL/HR)	13115	13115	45631	45631	45631	37565	8066	8066	37565	7985	45550	45550	714	3886	3886
H (MM BTU/HR)	-1338.9	-1310.9	-5488.1	-5434.5	-5320.2	-4379.8	-940.4	-904.6	-3866.7	-823.6	-4398	-4559	-72.4	-388.9	-372.1

STREAM	80	81	82	83	84	86	87	88	89	90	92	93	94	95	96
Temperature (F)	606.4	609.6	1050	1050.4	481.4	596	88.8	87.9	87.9	60	217.3	305.3	432.3	629.3	213
Pressure (PSIA)	350	350	342	342	35	60	0.7	0.7	40	14.7	16.3	72.5	352	1910.5	15
Flow (LB/HR)	750610	986887	986887	1056887	886460	110073	886460	886460	886460	194097	6828	130	2387	1453	3970
Flow (LBMOL/HR)	41665	54780	54780	58665	49205	6110	49205	49205	49205	10774	379	7	132	81	220
H (MM BTU/HR)	-4170.1	-5481	-5249.3	-5621.4	-4958.9	-609.8	-5182.8	-6040.7	-6040.6	-1328.1	-39	-0.9	-15.4	-9	-25.3

A-102

Humid Air Turbine (HAT)

Natural Gas / Pratt Whitney GT

Natural Gas HAT Cycle
(based on PW turbine)

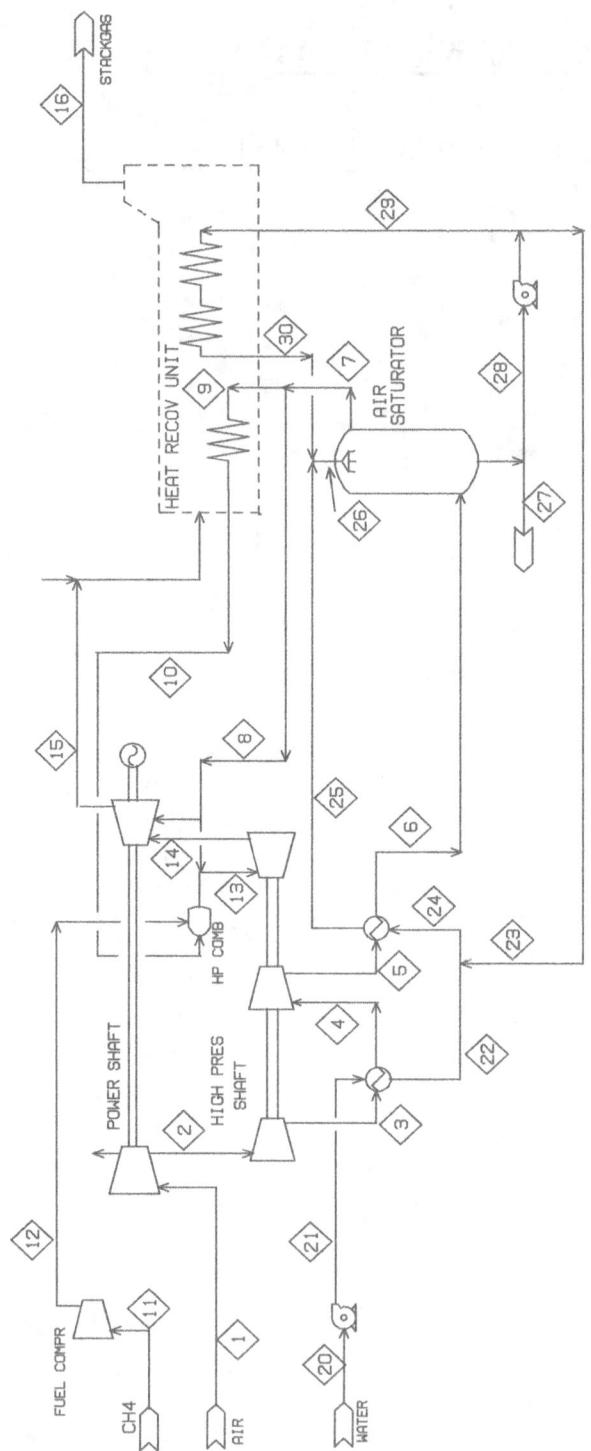

A-104

NATURAL GAS HAT CYCLE (based on PW turbine)

Gas Turb	326.5	MWe
Misc	3.0	MWe
Auxiliary	4.9	MWe
Net Power	318.7	MWe
Eff (HHV)	51.9	%
Eff (LHV)	57.6	%

STREAM ID	1	2	3	4	5	6	7	8	9	10	11	12	13	14
Temperature F	59	128.9	269.6	90	876.8	254	374.1	374.1	374.1	910	60	234.3	2750	2244.6
Pressure psi	14.54	21.81	43.62	39.62	796.9	765.3	742.4	742.4	742.4	727.5	250	780	691	275.13
Mass Flow lb/hr	2315881	2244960	2244960	2244960	2244960	2244960	2761017	335161	2425856	2425856	87727	87727	2679152	2679152
Mass Flow lb/sec	643.3	623.6	623.6	623.6	623.6	623.6	766.9	93.1	673.8	673.8	24.4	24.4	744.2	744.2
Mole Flow lbmol/hr	80268.5	77810.5	77810.5	77810.5	77810.5	77810.5	106463.7	12923.7	93540	93540	5468.3	5468.3	105392.6	105392.6
Enthalpy MMBtu/hr	-100.4	-59.5	17.1	-81	362.4	2.5	-2859	-347.1	-2512	-2110.4	-176.8	-168.9	-2450.7	-2978.6
Substream: MIXED														
Cp Btu/lb-R	0.242	0.242	0.244	0.243	0.263	0.257	0.31	0.31	0.31	0.314	0.555	0.641	0.395	0.382

STREAM ID	15	16	20	21	22	23	24	25	26	27	28	29	30
Temperature F	993.3	274.9	59	60.1	228	228.7	228.3	499.8	499.9	228.8	228.8	228.7	500
Pressure psi	15.49	15.2	14.7	815	798	804	798	783	783	744.4	744.4	804	783
Mass Flow lb/hr	2848744	2919664	540452	540452	540452	559761	1100213	1100213	1824139	24392	1283687	723926	723926
Mass Flow lb/sec	791.3	811	150.1	150.1	150.1	155.5	305.6	305.6	506.7	6.8	356.6	201.1	201.1
Mole Flow lbmol/hr	111932	114390.1	29999.6	29999.6	29999.6	31071.5	61071.1	61071.1	101255.1	1346.2	71255.5	40184	40184
Enthalpy MMBtu/hr	-4323.3	-4963.6	-3719.3	-3717.4	-3619.4	-3748.3	-7367.7	-7007.7	-11618.4	-161.1	-8595.8	-4847.6	-4610.7
Substream: MIXED													
Cp Btu/lb-R	0.33	0.297	1.078	1.076	1.097	1.098	1.097	1.422	1.423	1.086	1.098	1.098	1.423

Humid Air Turbine (HAT)

Coal Syngas / Destec (E-Gas^{TM}) / CGCU / Pratt Whitney GT

ASPEN IGHAT System

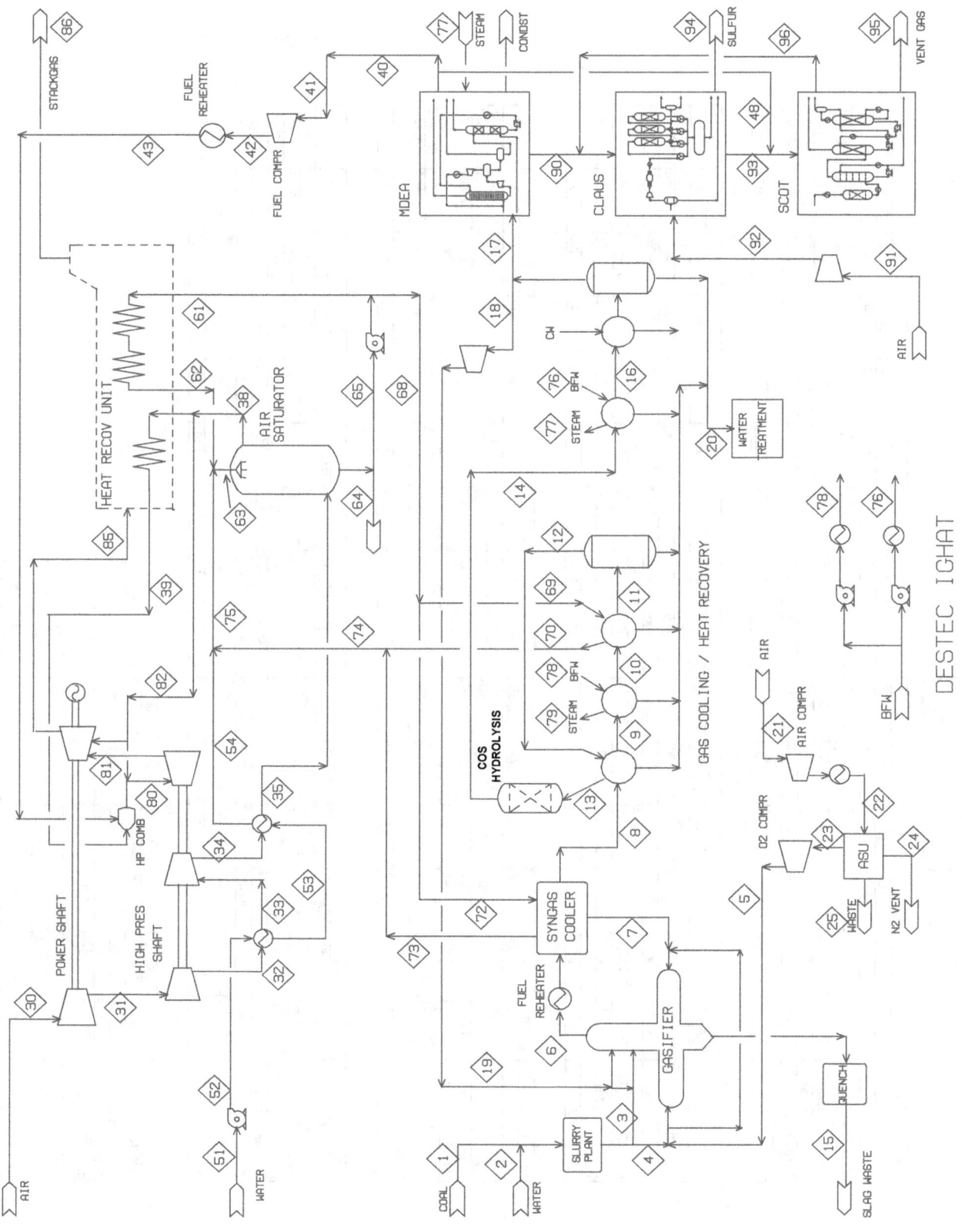

DESTEC IGHAT

A-107

IGHAT
Destec Gasifier
(slurry - 2 stage)

Gas Turb	457.6	MWe
Misc	44.0	MWe
Auxiliary	6.2	MWe
Net Power	407.3	MWe
Eff (HHV)	43.3	%
Eff (LHV)	44.9	%

Stream PFD ID	1	2	3	4	5	6	7	8	9	10	11	12	13	14	16
ASPEN ID	COALIN	WATERI	COALB	COALA	GASIFOXY	RXROUT	FINRCY	RAWGAZ	TOQR2	TOQR3	TOQR4	TOQR1A	TOCOS	TOQR5	TOQR7
Temperature F	59	59	350	350	291.4	1900	750	750	675	504.7	460	455	531.4	532	270
Pressure psi	14.7	14.7	465	465	464.1	412	418	407	402	397	392	387	382	380	370
Mass Flow lb/hr	275022	91639	76699	289961	208466	695583	10320	685263	685263	685263	685263	685263	685263	685263	685263
Mass Flow lb/sec	76.4	25.5	21.3	80.5	57.9	193.2	2.9	190.4	190.4	190.4	190.4	190.4	190.4	190.4	190.4
Mole Flow lbmol/hr		5086.8			6475.7			34766.7	34766.7	34766.7	34766.7	34766.7	34766.7	34766.7	34766.7
Enthalpy MMBtu/hr	-861	-630.7	-155	-583.5	9.4	-1108.1	-1.4	-1429.5	-1449.9	-1495.6	-1507.5	-1508.8	-1488.5	-1488.5	-1557.8
Substream: MIXED															
Cp Btu/lb-R	1.078	1.078			0.229	0.444		0.398	0.395	0.389	0.388	0.388	0.39	0.39	0.384

Stream PFD ID	17	18	19	20	21	22	23	24	25	30	31	32	33	34	35
ASPEN ID	TOCGCU	RECYGAS	GRCYC	TOSTRIP	AIRO2	ATASU2	GASO2A	VENTN2	WSTASU	GTAIR	AIRTHP	AIR1	AIR2	AIR6	AIR7
Temperature F	103	103	131.8	103.5	59	88	103	60	15	59	128.9	269.6	90	876.8	254
Pressure psi	365	365	425	20	14.54	20	16.5	16.5	15	14.54	21.81	43.62	39.62	796.9	785.7
Mass Flow lb/hr	485558	136952	136952	62754	898773	898773	208466	684876	5430	2315880	2244960	2244960	2244960	2244960	2244960
Mass Flow lb/sec	134.9	38	38	17.4	249.7	249.7	57.9	190.2	1.5	643.3	623.6	623.6	623.6	623.6	623.6
Mole Flow lbmol/hr	24420.2	6887.7	6887.8	3458.7	31151.5	31151.5	6475.7	24373.4	301.4	80268.5	77810.5	77810.5	77810.5	77810.5	77810.5
Enthalpy MMBtu/hr	-969.2	-273.4	-272	-423.7	-39	-29.9	-0.8	-2.3	-37.2	-100.4	-59.5	17.1	-81	362.4	2.4
Substream: MIXED															
Cp Btu/lb-R	0.371	0.371	0.373	1.064	0.242	0.244	0.244	0.216	0.248	1.077	0.242	0.244	0.243	0.263	0.257

Stream PFD ID	38	39	40	42	43	48	51	52	53	54	61	62	63	64	65
ASPEN ID	TOCO1	TOCOMB	FRSELEX	HPCPR	HPFUEL	REDGAS	WAT51	WAT52	WAT53	WAT54	WAT61	WAT62	WAT63	WAT64	WAT65
Temperature F	403.2	910	116	297	797.6	116	116	60.1	161.5	500	255.2	500	497.9	254.8	254.8
Pressure psi	742.4	727.5	319	780	757	319	14.7	815	798	783	804	783	783	744.4	744.4
Mass Flow lb/hr	3100276	2765227	448570	448570	448570	6755	899427	899427	899427	899427	1397524	1397524	3200708	43896	2301384
Mass Flow lb/sec	861.2	768.1	124.6	124.6	124.6	1.9	249.8	249.8	249.8	249.8	388.2	388.2	889.1	12.2	639.3
Mole Flow lbmol/hr	125307.8	111765.1	23240.4	23240.4	23240.4	350	49925.8	49925.8	49925.8	49925.8	77574.4	77574.4	177666.3	2418.9	127746.2
Enthalpy MMBtu/hr	-4760.7	-3777.6	-884.4	-854.2	-766.2	-13.3	-6189.7	-6186.6	-6088.6	-5728.5	-9317.3	-8900.9	-20395.2	-287.6	-15344.6
Substream: MIXED															
Cp Btu/lb-R	0.337	0.338	0.378	0.388	0.398	0.378	1.078	1.076	1.08	1.423	1.108	1.423	1.416	1.093	1.108

Stream PFD ID	68	69	70	72	73	74	75	76	77	78	79	80	81	82	85
ASPEN ID	WAT68	WAT69	WAT70	WAT72	WAT73	WAT74	WAT75	HTWCG	SHSCG	HTWUT	SHS115	POCX	POC3	ABLEED	GTPOC
Temperature F	255.2	255.2	397.8	255.2	500	492.4	496.2	250	265	250	375	2750.2	2364.4	403.2	1034.6
Pressure psi	804	804	783	804	783	783	783	783	35	165	160	691	335.53	742.4	15.2
Mass Flow lb/hr	903757	72301	72301	831457	831457	903757	1803184	71217	71217	43895	43895	3379367	3379367	335161	3619879
Mass Flow lb/sec	251	20.1	20.1	231	231	251	500.9	19.8	19.8	12.2	12.2	938.7	938.7	93.1	1005.5
Mole Flow lbmol/hr	50166.1	4013.3	4013.3	46152.9	46152.9	50166.1	100091.9	3953.2	3953.2	2436.5	2436.5	131289.7	131289.7	13546.6	140602.4
Enthalpy MMBtu/hr	-6025.3	-482	-470.1	-5543.3	-5295.6	-5765.7	-11494.3	-475.3	-405.2	-293	-248	-4798.1	-5326	-514.6	-7210.7
Substream: MIXED															
Cp Btu/lb-R	1.108	1.108	1.219	1.108	1.423	1.4	1.411	1.11	0.461	1.109	0.49	0.408	0.398	0.337	0.34

Stream PFD ID	86	90	91	92	93	94	95	96
ASPEN ID	STACK	AG-CLUS1	HPCAIR	AIRTCL	CL-TAIL1	CLAUSULF	TG-SCOT	TG-SCOTR
Temperature F	273.2	141.1	59	171.7	439.2	285	70	70
Pressure psi	14.8	18.5	14.7	25	26.7	14.7	17.5	17.5
Mass Flow lb/hr	3619879	33506	15161	15161	43849	6802	48620	1985
Mass Flow lb/sec	1005.5	9.3	4.2	4.2	12.2	1.9	13.5	0.6
Mole Flow lbmol/hr	140602.4	1011.5	527.4	527.4	1478.3	212.1	1779.5	46.5
Enthalpy MMBtu/hr	-8095.5	-87.5	-0.7	-0.2	-111.8	-0.7	-129.9	-6.9
Substream: MIXED								
Cp Btu/lb-R	0.303	0.248	0.243	0.244	0.282	0.359	0.207	

Appendix B

Cost of Electricity (COE) Analysis

Cost of Electricity Analysis

The cost of electricity was evaluated using data from the EG&G Cost Estimating notebook (version 1.11) and several contractor reports. The format follows the guidelines set by EPRI TAG. The individual section costs for each case are listed in the following COE spreadsheet summarizies and are based on capacity-factored techniques. All costs are reported in 1st Quarter 2002 dollars.

.

Bulk Plant Items

Bulk plant items include water systems, civil/structural/architectural, piping, control and instrumentation, and electrical systems. These were calculated based on a percentage of the total installed equipment costs. The percentages in parenthesis, for coal systems, are for the hot-gas cleanup process, which has a lower water requirement, and therefore, a smaller percentage for piping and water systems. The following percentages were used in this report.

	% of Installed Equipment Cost		
Plant Type :	**Natural Gas**	**PC Plant**	**Coal**
Bulk Plant Item			
Water Systems	7.1	6.3	5.5 (3.5)
Civil/Structural/Architectural	13.9	10.0	6.2
Piping	7.1	6.3	5.5 (3.5)
Control and Instrumentation	8.0	6.0	4.0
Electrical Systems	15.8	12.2	8.7
Total	51.9	40.8	29.9 (25.9)

Table 1, Table 2, and Table 3 show the assumptions used in this COE analysis.

Table 1.   Capital Cost Assumptions

Engineering Fee	10% of PPC*
Project Contingency	15% of PPC
Construction Period	4 Yrs (coal), 2 Yrs (NG)
Inflation Rate	3%
Discount Rate	11.2%
Prepaid Royalties	0.5% of PPC
Catalyst and Chemical Inventory	30 Dys
Spare Parts	0.5% of TPC**
Land	200 Acres @ $6,500/Acre

Start-Up Costs	
Plant Modifications	2% of TPI***
Operating Costs	30 Dys
Fuel Costs	7.5 Dys

Working Capital	
Coal	60 Dys
By-Product Inventory	30 Dys
O&M Costs	30 Dys

*   PPC = Process Plant Cost
**   TPC = Total Plant Cost
***   TPI = Total Plant Investment

Table 2. Operating & Maintenance Assumptions

Consumable Material Prices

Illinois #6 Coal	$24.36/Ton
Natural Gas	$3.20 / 1000 SCF
Raw Water	$0.19 /Ton
MDEA Solvent	$1.45/Lb
Claus Catalyst	$470/Ton
SCOT Activated Alumina	$0.067/Lb
Sorbent	$6,000/Ton
Nahcolite	$275/Ton
Limestone (FGD)	$16/Ton
Off-Site Ash/Sorbent Disposal Costs	$8.00/Ton
Operating Royalties	1% of Fuel Cost
Operator Labor	$34.00/hour
Number of Shifts for Continuous Operation	4.2
Supervision and Clerical Labor	30% of O&M Labor
Maintenance Costs	2.2% of TPC
Insurance and Local Taxes	2% of TPC
Miscellaneous Operating Costs	10% of O&M Labor
Capacity Factor	85%

Table 3. Investment Factor Economic Assumptions

Annual Inflation Rate		3%	
Real Escalation Rate (over inflation)			
O&M		0%	
Coal		-1.1%	
Natural Gas		0.2%	
Discount Rate		11.2%	
Debt	80% of Total	9.0% Cost	7.2% Return
Preferred Stock	0% of Total	0.0% Cost	0% Return
Common Stock	20% of Total	20.0% Cost	4.0% Return
			11.2% Total
Book Life		20 Yrs	
Tax Life		20 Yrs	
State and Federal Tax Rate		38%	
Investment Tax Credit		0%	
Number of Years Levelized Cost		10 Yrs	

Cost of Electricity (COE) Spreadsheet Listings

Case	Page

Pulverized Coal (PC)

PC Steam Cycle – No CO_2 Capture

PULIVERIZED COAL (PC) PLANT

397 MW POWER PLANT

1st Q 2002 Dollar

Total Plant Investment		PROCESS	PROCESS	COST, K$
AREA NO	PLANT SECTION DESCRIPTION	CONT, %	CONT, K$	W/O CONT
11	Coal Preparation & Feed	0	$0	$17,617
12	PC Boiler, Fans & Accessories	0	$0	$75,094
13	Flue Gas Cleanup (Precipitator,FGD)	0	$0	$56,290
13	Sorbent Preparation & Handling	0	$0	$6,002
13	Stack / Ductwork	0	$0	$18,816
15	Steam Turbine & Accessories	0	$0	$59,088
14	Spent Sorbent/Ash disposal system	0	$0	$18,273
18	Water Systems	0	$0	$15,824
30	Civil/Structural/Architectural	0	$0	$25,118
40	Piping	0	$0	$15,824
50	Control/ Instrumentation	0	$0	$15,071
60	Electrical	0	$0	$30,644

Subtotal, Process Plant Cost	$353,660

Engineering Fees		$35,366
Process Contingency (Using cont. listed)		$0
Project Contingency,	15 % Proc Plt & Gen Plt Fac	$53,049

Total Plant Cost (TPC)	$442,075

Plant Construction Period,	3.0 Years (1 or more)	
Construction Interest Rate,	11.2 %	
Adjustment for Interest and Inflation		$36,030

Total Plant Investment (TPI)	$478,105

Prepaid Royalties		$1,768
Initial Catalyst and Chemical Inventory		$333
Startup Costs		$12,273
Spare Parts		$2,210
Working Capital		$7,103
Land,	200 Acres	$1,300

Total Capital Requirement (TCR)		$503,092
	$/kW	1268

ANNUAL OPERATING COSTS

Capacity Factor = 85 %

COST ITEM	QUANTITY	UNIT $ PRICE	ANNUAL COST, K$
Coal (Illinois #6)	3714 T/D	$24.36 /T	$28,066
Consumable Materials			
Water	38,160 T/D	$0.19 /T	$2,249
Limestone	363.0 T/D	$16.00 /T	$1,802
Ash/Sorbent Disposal Costs	739 T/D	$8.00 /T	$1,835
Plant Labor			
Oper Labor (incl benef)	15 Men/shift	$34.00 /Hr.	$4,455
Supervision & Clerical			$2,504
Maintenance Costs	2.2%		$9,726
Royalties			$281
Other Operating Costs			$835
Total Operating Costs			$51,752
By-Product Credits			
	0.0 T/D	$0.00 /T	$0
	0.0 T/D	$0.00 /T	$0
Total By-Product Credits			$0
Net Operating Costs			$51,752

BASES AND ASSUMPTIONS

A. CAPITAL BASES AND DETAILS

	QUANTITY	UNIT $ PRICE	COST, K$
Initial Cat./Chem. Inventory			
Water	973080 T	$0.19 /T	$185
Limestone	9257 T	$16.00 /T	$148
Total Catalyst and Chemical Inventory			$333
Startup costs			
Plant modifications,	2 % TPI		$9,562
Operating costs			$2,032
Fuel			$678
Total Startup Costs			$12,273
Working capital			
Fuel & Consumables inv	60 days supply		$6,211
By-Product inventory	30 days supply		$0
Direct expenses	30 days		$892
Total Working Capital			$7,103

B. ECONOMIC ASSUMPTIONS

Project life	20 Years
Book life	20 Years
Tax life	20 Years
Federal and state income tax rate	38.0 %
Tax depreciation method	ACRS
Investment Tax Credit	0.0 %

Financial structure

Type of Security	% of Total	Current Dollar Cost, %	Ret, %	Constant Dollar Cost, %	Ret, %
Debt	80	9.0	7.2	5.8	4.6
Preferred Stock	0	3.0	0.0	0.0	0.0
Common Stock	20	20.0	4.0	16.5	3.3
Discount rate (cost of capital)		11.2		7.9	

Inflation rate, % per year	3.0
Real Escalation rates (over inflation)	
Fuel, % per year	-1.1
Operating & Maintenance, % per year	0.0

C. COST OF ELECTRICITY

The approach to determining the cost of electricity is based upon the methodology described in the Technical Assessment Guide, published by the Electric Power Research Institute. The cost of electricity is stated in terms of 10th year levelized dollars.

	Current $	Constant $
Levelizing Factors		
Capital Carrying Charge, 10th yr	0.179	0.148
Fuel, 10th year	1.091	0.948
Operating & Maintenance, 10th yr	1.151	1.000

Cost of Electricity - Levelized	mills/kWh	mills/kWh
Capital Charges	30.5	25.3
Fuel Costs	10.4	9.0
Consumables	2.3	2.0
Fixed Operating & Maintenance	5.9	5.1
Variable Operating & Maintenance	1.0	0.9
By-product	0.0	0.0
Total Cost of Electricity	50.1	42.3

Pulverized Coal (PC)

PC Steam Cycle - Amine CO_2 Capture

PULIVERIZED COAL (PC) PLANT
AMINE CASE

283 MW POWER PLANT

1st Q 2002 Dollar

AREA NO	PLANT SECTION DESCRIPTION	PROCESS CONT, %	PROCESS CONT, K$	COST, K$ W/O CONT
11	Coal Preparation & Feed	0	$0	$17,617
12	PC Boiler, Fans & Accessories	0	$0	$75,094
13	Flue Gas Cleanup (Precipitator,FGD)	0	$0	$56,290
13	Sorbent Preparation & Handling	0	$0	$6,002
13	Stack / Ductwork	0	$0	$15,664
15	Steam Turbine & Accessories	0	$0	$50,898
14	Spent Sorbent/Ash disposal system	0	$0	$18,273
15	Amine Plant	0	$0	$92,423
16	CO2 Compression	0	$0	$30,103
18	Water Systems	0	$0	$17,006
30	Civil/Structural/Architectural	0	$0	$26,994
40	Piping	0	$0	$17,006
50	Control/ Instrumentation	0	$0	$16,196
60	Electrical	0	$0	$32,933

Subtotal, Process Plant Cost		$472,500

Engineering Fees		$47,250
Process Contingency (Using cont. listed)		$0
Project Contingency,	15 % Proc Plt & Gen Plt Fac	$70,875

Total Plant Cost (TPC)		$590,625

Plant Construction Period,	3.0 Years (1 or more)	
Construction Interest Rate,	11.2 %	
Adjustment for Interest and Inflation		$48,137

Total Plant Investment (TPI)		$638,761

Prepaid Royalties		$2,362
Initial Catalyst and Chemical Inventory		$969
Startup Costs		$16,538
Spare Parts		$2,953
Working Capital		$8,739
Land,	200 Acres	$1,300

Total Capital Requirement (TCR)		$671,624
	$/kW	2373

ANNUAL OPERATING COSTS

Capacity Factor =   85 %

COST ITEM	QUANTITY		UNIT $ PRICE	ANNUAL COST, K$
Coal (Illinois #6)	3714	T/D	$24.36 /T	$28,066
Consumable Materials				
Water	38,160	T/D	$0.19 /T	$2,249
Limestone	363	T/D	$16.00 /T	$1,802
Amine Chemicals	8,315	T/D	$3.00 /T CO2 captur	$7,739
Ash/Sorbent Disposal Costs	707	T/D	$8.00 /T	$1,756
Plant Labor				
Oper Labor (incl benef)	15	Men/shift	$34.00 /Hr.	$4,455
Supervision & Clerical				$2,896
Maintenance Costs	2.2%			$12,994
Royalties				$281
Other Operating Costs				$965
Total Operating Costs				**$63,203**
By-Product Credits				
	0.0	T/D	$0.00 /T	$0
	0.0	T/D	$0.00 /T	$0
Total By-Product Credits				$0
Net Operating Costs				**$63,203**

BASES AND ASSUMPTIONS

A. CAPITAL BASES AND DETAILS

	QUANTITY	UNIT $ PRICE	COST, K$
Initial Cat./Chem. Inventory			
Water	973080 T	$0.19 /T	$185
Limestone	9257 T	$16.00 /T	$148
Amine Chemicals	212033 T	$3.00 /T	$636
Total Catalyst and Chemical Inventory			$969
Startup costs			
Plant modifications,	2 % TPI		$12,775
Operating costs			$3,084
Fuel			$678
Total Startup Costs			$16,538
Working capital			
Fuel & Consumables inv	60 days supply		$7,708
By-Product inventory	30 days supply		$0
Direct expenses	30 days		$1,031
Total Working Capital			$8,739

B. ECONOMIC ASSUMPTIONS

Project life	20 Years
Book life	20 Years
Tax life	20 Years
Federal and state income tax rate	38.0 %
Tax depreciation method	ACRS
Investment Tax Credit	0.0 %

Financial structure

Type of Security	% of Total	Current Dollar Cost, %	Ret, %	Constant Dollar Cost, %	Ret, %
Debt	80	9.0	7.2	5.8	4.6
Preferred Stock	0	3.0	0.0	0.0	0.0
Common Stock	20	20.0	4.0	16.5	3.3
Discount rate (cost of capital)			11.2		7.9

Inflation rate, % per year	3.0
Real Escalation rates (over inflation)	
Fuel, % per year	-1.1
Operating & Maintenance, % per year	0.0

B-15

C. COST OF ELECTRICITY

The approach to determining the cost of electricity is based upon the methodology described in the Technical Assessment Guide, published by the Electric Power Research Institute. The cost of electricity is stated in terms of 10th year levelized dollars.

	Current $	Constant $
Levelizing Factors		
Capital Carrying Charge, 10th yr	0.179	0.148
Fuel, 10th year	1.091	0.948
Operating & Maintenance, 10th yr	1.151	1.000

Cost of Electricity - Levelized	mills/kWh	mills/kWh
Capital Charges	57.1	47.3
Fuel Costs	14.5	12.6
Consumables	7.4	6.4
Fixed Operating & Maintenance	10.0	8.7
Variable Operating & Maintenance	1.8	1.5
By-product	0.0	0.0
Total Cost of Electricity	90.8	76.6

B-16

Pulverized Coal (PC)

PC Steam Cycle - O_2 Boiler / CO_2 Capture

PULIVERIZED COAL (PC) PLANT
CRYOGENIC CASE

298 MW POWER PLANT

1st Q 2002 Dollar

Total Plant Investment

AREA NO	PLANT SECTION DESCRIPTION	PROCESS CONT, %	PROCESS CONT, K$	COST, K$ W/O CONT
11	Coal Preparation & Feed	0	$0	$17,118
12	PC Boiler, Fans & Accessories	0	$0	$72,808
13	Flue Gas Cleanup (Precipitator,FGD)	0	$0	$51,632
13	Sorbent Preparation & Handling	0	$0	$6,002
13	Stack / Ductwork	0	$0	$1,009
15	Steam Turbine & Accessories	0	$0	$58,828
14	Spent Sorbent/Ash disposal system	0	$0	$17,882
15	Oxygen Plant	0	$0	$111,099
16	CO2 Compression	0	$0	$34,208
18	Water Systems	0	$0	$16,348
30	Civil/Structural/Architectural	0	$0	$25,949
40	Piping	0	$0	$16,348
50	Control/ Instrumentation	0	$0	$15,569
60	Electrical	0	$0	$31,657

Subtotal, Process Plant Cost				$476,456

Engineering Fees				$47,646
Process Contingency (Using cont. listed)				$0
Project Contingency,	15 % Proc Plt & Gen Plt Fac			$71,468

Total Plant Cost (TPC)				$595,570

Plant Construction Period,	3.0 Years (1 or more)			
Construction Interest Rate,	11.2 %			
Adjustment for Interest and Inflation				$48,540

Total Plant Investment (TPI)				$644,110

Prepaid Royalties				$2,382
Initial Catalyst and Chemical Inventory				$327
Startup Costs				$15,869
Spare Parts				$2,978
Working Capital				$6,998
Land,	200 Acres			$1,300

Total Capital Requirement (TCR)				$673,964
			$/kW	2259

ANNUAL OPERATING COSTS

Capacity Factor =   85 %

COST ITEM		QUANTITY	UNIT $ PRICE	ANNUAL COST, K$
Coal (Illinois #6)		3553 T/D	$24.36 /T	$26,854
Consumable Materials				
Water		38,160 T/D	$0.19 /T	$2,249
Limestone		347 T/D	$16.00 /T	$1,724
Ash/Sorbent Disposal Costs		707 T/D	$8.00 /T	$1,756
Plant Labor				
Oper Labor (incl benef)		15 Men/shift	$34.00 /Hr.	$4,455
Supervision & Clerical				$2,909
Maintenance Costs	2.2%			$13,103
Royalties				$269
Other Operating Costs				$970
Total Operating Costs				$54,288
By-Product Credits				
		0.0 T/D	$0.00 /T	$0
		0.0 T/D	$0.00 /T	$0
Total By-Product Credits				$0
Net Operating Costs				$54,288

BASES AND ASSUMPTIONS

A. CAPITAL BASES AND DETAILS

	QUANTITY	UNIT $ PRICE	COST, K$
Initial Cat./Chem. Inventory			
Water	973080 T	$0.19 /T	$185
Limestone	8858 T	$16.00 T	$142
Total Catalyst and Chemical Inventory			$327
Startup costs			
Plant modifications,	2 % TPI		$12,882
Operating costs			$2,338
Fuel			$649
Total Startup Costs			$15,869
Working capital			
Fuel & Consumables inv	60 days supply		$5,962
By-Product inventory	30 days supply		$0
Direct expenses	30 days		$1,036
Total Working Capital			$6,998

B. ECONOMIC ASSUMPTIONS

Project life	20 Years
Book life	20 Years
Tax life	20 Years
Federal and state income tax rate	38.0 %
Tax depreciation method	ACRS
Investment Tax Credit	0.0 %

Financial structure

Type of Security	% of Total	Current Dollar Cost, %	Ret, %	Constant Dollar Cost, %	Ret, %
Debt	80	9.0	7.2	5.8	4.6
Preferred Stock	0	3.0	0.0	0.0	0.0
Common Stock	20	20.0	4.0	16.5	3.3
Discount rate (cost of capital)			11.2		7.9

Inflation rate, % per year	3.0
Real Escalation rates (over inflation)	
Fuel, % per year	-1.1
Operating & Maintenance, % per year	0.0

C. COST OF ELECTRICITY

The approach to determining the cost of electricity is based upon the
methodology described in the Technical Assessment Guide, published by
the Electric Power Research Institute. The cost of electricity is
stated in terms of 10th year levelized dollars.

	Current $	Constant $
Levelizing Factors		
Capital Carrying Charge, 10th yr	0.179	0.148
Fuel, 10th year	1.091	0.948
Operating & Maintenance, 10th yr	1.151	1.000

Cost of Electricity - Levelized	mills/kWh	mills/kWh
Capital Charges	54.3	45.0
Fuel Costs	13.2	11.5
Consumables	3.0	2.6
Fixed Operating & Maintenance	9.6	8.3
Variable Operating & Maintenance	1.7	1.5
By-product	0.0	0.0
Total Cost of Electricity	81.7	68.8

Combined Cycle

Natural Gas Combined Cycle (NGCC) - No CO_2 Capture

Natural Gas Combined Cycle
W501G

379 MW POWER PLANT

Total Plant Investment

1st Q 2002 Dollar

AREA NO	PLANT SECTION DESCRIPTION	PROCESS CONT, %	PROCESS CONT, K$	COST, K$ W/O CONT
15	Gas Turbine	5	$2,619	$52,388
15	Steam Cycle	5	$2,103	$42,065
18	Water Systems	0		$6,706
30	Civil/Structural/Architectural	0		$13,129
40	Piping	0		$6,706
50	Control/ Instrumentation	0		$7,556
60	Electrical	0		$14,924
	Subtotal, Process Plant Cost			**$128,551**

Engineering Fees			$12,855
Process Contingency (Using cont. listed)			$4,723
Project Contingency,	15 % Proc Plt & Gen Plt Fac		$19,283

Total Plant Cost (TPC)		**$165,412**
	$/kw	$436

Plant Construction Period,	2.0 Years (1 or more)	
Construction Interest Rate,	11.2 %	
Adjustment for Interest and Inflation		$6,567

Total Plant Investment (TPI)	**$171,979**

Prepaid Royalties			$643
Initial Catalyst and Chemical Inventory			$5
Startup Costs			$9,925
Spare Parts			$827
Working Capital			$11,705
Land,	100 Acres	@ $1500/acre	$150

Total Capital Requirement (TCR)		**$195,233**
	$/kW	515

ANNUAL OPERATING COSTS

Capacity Factor =   85 %

Consumables			UNIT $	ANNUAL
COST ITEM	QUANTITY		PRICE	COST, K$
Natural Gas	58,760	1000 SCF/day	$3.20 /1000 SCF	$58,337
Water	2,263	1000 gal/day	$0.80 /1000 gal	$562
Plant Labor				
Oper Labor (incl benef)		5 Men/shift	$34.00 /Hr.	$1,485
Supervision & Clerical				$882
Maintenance Costs	2.2%			$3,639
Insurance & Local Taxes				$3,308
Other Operating Costs				$294
Total Operating Costs				$68,507

BASES AND ASSUMPTIONS

A. CAPITAL BASES AND DETAILS

	QUANTITY		UNIT $ PRICE	COST, K$
Initial Cat./Chem. Inventory				
Water	5,771	1000 gallons	$0.80 /1000 gal	$5
Total Catalyst and Chemical Inventory				$5
Startup costs				
Plant modifications,	2	% TPI		$3,440
Operating costs				$6,485
Total Startup Costs				$9,925
Working capital				
Fuel & Consumables inv	60	days supply		$11,391
Direct expenses	30	days		$314
Total Working Capital				$11,705

B. ECONOMIC ASSUMPTIONS

Project life	20 Years
Book life	20 Years
Tax life	20 Years
Federal and state income tax rate	38.0 %
Tax depreciation method	MACRS
Investment Tax Credit	0.0 %
Financial structure	

Type of Security	% of Total	Current Dollar Cost, %	Ret, %	Constant Dollar Cost, %	Ret, %
Debt	80	9.0	7.2	5.8	4.6
Preferred Stock	0	0.0	0.0	0.0	0.0
Common Stock	20	20.0	4.0	16.5	3.3
Discount rate (cost of capital)			11.2		7.9

Inflation rate, % per year	3.0
Real Escalation rates (over inflation)	
Fuel, % per year	0.2
Operating & Maintenance, % per year	0.0

C. COST OF ELECTRICITY

The approach to determining the cost of electricity is based upon the methodology described in the Technical Assessment Guide, published by the Electric Power Research Institute. The cost of electricity is stated in terms of 10th year levelized dollars.

	Current $	Constant $
Levelizing Factors		
Capital Carrying Charge, 10th yr	0.179	0.148
Fuel, 10th year	1.162	1.010
Operating & Maintenance, 10th yr	1.151	1.000

Cost of Electricity - Levelized	mills/kWh	mills/kWh
Capital Charges	12.4	10.3
Fuel Costs	24.0	20.9
Consumables	0.2	0.2
Fixed Operating & Maintenance	3.3	2.9
Variable Operating & Maintenance	0.6	0.5
By-product	0.0	0.0
Total Cost of Electricity	**40.5**	**34.7**

Combined Cycle

Natural Gas Combined Cycle (NGCC) - CO_2 Capture

Natural Gas Combined Cycle　　　　　　　　　　　　　**327 MW POWER PLANT**

W501G + CO2 CAPTURE　　　　　　　　　　　　　　　　1st Q 2002 Dollar

Total Plant Investment

AREA NO	PLANT SECTION DESCRIPTION	PROCESS CONT, %	PROCESS CONT, K$	COST, K$ W/O CONT
15	Gas Turbine	5	$2,619	$52,389
15	Steam Cycle	5	$1,872	$37,438
20	Amine System	5	$3,453	$69,053
20	CO2 Compression/drying	5	$526	$10,530
18	Water Systems	0		$6,378
30	Civil/Structural/Architectural	0		$12,486
40	Piping	0		$6,378
50	Control/ Instrumentation	0		$7,186
60	Electrical	0		$14,193
	Subtotal, Process Plant Cost			$201,836

Engineering Fees		$20,184
Process Contingency (Using cont. listed)		$8,470
Project Contingency,	15 % Proc Plt & Gen Plt Fac	$30,275

	Total Plant Cost (TPC)		$260,765
		$/kw	$798

Plant Construction Period,	2.0 Years (1 or more)	
Construction Interest Rate,	11.2 %	
Adjustment for Interest and Inflation		$10,352

Total Plant Investment (TPI)		$271,117

Prepaid Royalties			$1,009
Initial Catalyst and Chemical Inventory			$5
Startup Costs			$12,264
Spare Parts			$1,304
Working Capital			$11,794
Land,	100 Acres	@ $1500/acre	$150

Total Capital Requirement (TCR)		$297,643
	$/kW	911

ANNUAL OPERATING COSTS

Capacity Factor =   85 %

| Consumables | | | UNIT $ | ANNUAL |
COST ITEM	QUANTITY		PRICE	COST, K$
Natural Gas	58,760	1000 SCF/day	$3.20 /1000 SCF	$58,337
Water	2,263	1000 gal/day	$0.80 /1000 gallons	$562
Amine Chemicals	130	ton CO2/hr	$3.00 /ton CO2 Capt	$2,893

Plant Labor				
Oper Labor (incl benef)		5 Men/shift	$34.00 /Hr.	$1,485
Supervision & Clerical				$1,134

Maintenance Costs	2.2%			$5,737

Insurance & Local Taxes				$5,215

Other Operating Costs				$378

Total Operating Costs				$72,848

BASES AND ASSUMPTIONS

A. CAPITAL BASES AND DETAILS

	QUANTITY		UNIT $ PRICE		COST, K$
Initial Cat./Chem. Inventory					
Water	5,771	1000 gallons	$0.80	/1000 gallons	$5
Amine Chemicals	18,492	(7 days CO2)	$3.00	/ton CO2	$55
Total Catalyst and Chemical Inventory					$5
Startup costs					
Plant modifications,	2	% TPI			$5,422
Operating costs					$6,842
Total Startup Costs					$12,264
Working capital					
Fuel & Consumables inv	60	days supply			$11,391
Direct expenses	30	days			$404
Total Working Capital					$11,794

B. ECONOMIC ASSUMPTIONS

Project life	20 Years
Book life	20 Years
Tax life	20 Years
Federal and state income tax rate	38.0 %
Tax depreciation method	MACRS
Investment Tax Credit	0.0 %
Financial structure	

Type of Security	% of Total	Current Dollar		Constant Dollar	
		Cost, %	Ret, %	Cost, %	Ret, %
Debt	80	9.0	7.2	5.8	4.6
Preferred Stock	0	0.0	0.0	0.0	0.0
Common Stock	20	20.0	4.0	16.5	3.3
Discount rate (cost of capital)		11.2		7.9	

Inflation rate, % per year	3.0
Real Escalation rates (over inflation)	
Fuel, % per year	0.2
Operating & Maintenance, % per year	0.0

C. COST OF ELECTRICITY

The approach to determining the cost of electricity is based upon the methodology described
in the Technical Assessment Guide, published by the Electric Power Research Institute.
The cost of electricity is stated in terms of 10th year levelized dollars.

	Current $	Constant $
Levelizing Factors		
Capital Carrying Charge, 10th yr	0.179	0.148
Fuel, 10th year	1.162	1.010
Operating & Maintenance, 10th yr	1.151	1.000

Cost of Electricity - Levelized	mills/kWh	mills/kWh
Capital Charges	21.9	18.2
Fuel Costs	27.9	24.2
Consumables	0.3	0.2
Fixed Operating & Maintenance	5.6	4.9
Variable Operating & Maintenance	1.0	0.9
By-product	0.0	0.0
Total Cost of Electricity	56.6	48.3

Combined Cycle

IGCC Destec (E-Gas^{TM}) / CGCU / "G" Gas Turbine

Destec CGCU IGCC   **401 MW POWER PLANT**

1st Q 2002 Dollar

Total Plant Investment

AREA NO	PLANT SECTION DESCRIPTION	PROCESS CONT, %	PROCESS CONT, K$	COST, K$ W/O CONT
11	Coal Slurry Preparation	0	$0	$27,007
12	Oxygen Plant	0	$0	$49,777
12	Destec Gasifier	5	$3,144	$62,876
12	Recycle Gas Compressor	5	$135	$2,696
14	Low Temperature Gas Cooling	0	$0	$13,986
14	MDEA	0	$0	$5,298
14	Claus	0	$0	$10,129
14	SCOT	0	$0	$4,284
15	Gas Turbine System	5	$2,871	$57,410
15	HRSG/Steam Turbine	5	$2,463	$49,269
18	Water Systems	0	$0	$15,550
30	Civil/Structural/Architectural	0	$0	$17,529
40	Piping	0	$0	$15,550
50	Control/ Instrumentation	0	$0	$11,309
60	Electrical	0	$0	$24,598

Subtotal, Process Plant Cost		$367,266

Engineering Fees		$36,727
Process Contingency (Using cont. listed)		$8,613
Project Contingency,	15 % Proc Plt & Gen Plt Fac	$55,090

Total Plant Cost (TPC)	$467,695

Plant Construction Period,	4.0 Years (1 or more)	
Construction Interest Rate,	11.2 %	
Adjustment for Interest and Inflation		$58,710

Total Plant Investment (TPI)	$526,405

Prepaid Royalties		$1,836
Initial Catalyst and Chemical Inventory		$69
Startup Costs		$12,745
Spare Parts		$2,338
Working Capital		$5,719
Land,	200 Acres	$1,300

Total Capital Requirement (TCR)		$550,414
	$/kW	1374

ANNUAL OPERATING COSTS

Capacity Factor =   85 %

COST ITEM	QUANTITY	UNIT $ PRICE	ANNUAL COST, K$
Coal (Illinois #6)	3,123 T/D	$24.36 /T	$23,600
Consumable Materials			
Water	2,924 T/D	$0.19 /T	$172
MDEA Solvent	403.2 Lb/D	$1.45 /Lb	$181
Claus Catalyst	0.01 T/D	$470 /T	$1
SCOT Activated Alumin	15.9 Lb/D	$0.67 /Lb	$3
SCOT Cobalt Catalyst			$5
SCOT Chemicals			$16
Ash/Sorbent Disposal Costs	413 T/D	$8.00 /T	$1,024
Plant Labor			
Oper Labor (incl benef)	15 Men/shift	$34.00 /Hr.	$4,455
Supervision & Clerical			$2,571
Maintenance Costs	2.2%		$10,289
Royalties			$236
Other Operating Costs			$857
Total Operating Costs			**$43,412**
By-Product Credits			
Sulfur	75.5 T/D	$75.00 /T	$1,757
	0.0 T/D	$0.00 /T	$0
Total By-Product Credits			$1,757
Net Operating Costs			**$41,655**

BASES AND ASSUMPTIONS

A. CAPITAL BASES AND DETAILS

	QUANTITY	UNIT $ PRICE	COST, K$
Initial Cat./Chem. Inventory			
Water	74550 T	$0.19 /T	$14
MDEA Solvent	10282 Lb	$1.45 /Lb	$15
Claus Catalyst	0.3 T	$470 /T	$0
SCOT Activated Alumin	405 Lb	$0.67 /Lb	$0
SCOT Cobalt Catalyst			$16
SCOT Chemicals			$24
Total Catalyst and Chemical Inventory			$69
Startup costs			
Plant modifications,	2 % TPI		$10,528
Operating costs			$1,647
Fuel			$570
Total Startup Costs			$12,745
Working capital			
Fuel & Consumables inv	60 days supply		$4,633
By-Product inventory	30 days supply		$170
Direct expenses	30 days		$916
Total Working Capital			$5,719

B. ECONOMIC ASSUMPTIONS

Project life	20 Years
Book life	20 Years
Tax life	20 Years
Federal and state income tax rate	38.0 %
Tax depreciation method	MACRS
Investment Tax Credit	0.0 %
Financial structure	

Type of Security	% of Total	Current Dollar Cost, %	Ret, %	Constant Dollar Cost, %	Ret, %
Debt	80	9.0	7.2	5.8	4.6
Preferred Stock	0	3.0	0.0	0.0	0.0
Common Stock	20	20.0	4.0	16.5	3.3
Discount rate (cost of capital)			11.2		7.9

Inflation rate, % per year	3.0
Real Escalation rates (over inflation)	
Fuel, % per year	-1.1
Operating & Maintenance, % per year	0.0

B-35

C. COST OF ELECTRICITY

The approach to determining the cost of electricity is based upon the methodology described in the Technical Assessment Guide, published by the Electric Power Research Institute. The cost of electricity is stated in terms of 10th year levelized dollars.

	Current $	Constant $
Levelizing Factors		
Capital Carrying Charge, 10th yr	0.179	0.148
Fuel, 10th year	1.091	0.948
Operating & Maintenance, 10th yr	1.151	1.000

Cost of Electricity - Levelized	mills/kWh	mills/kWh
Capital Charges	33.0	27.4
Fuel Costs	8.6	7.5
Consumables	0.5	0.5
Fixed Operating & Maintenance	6.0	5.2
Variable Operating & Maintenance	1.1	0.9
By-product	-0.7	-0.6
Total Cost of Electricity	48.6	40.9

Combined Cycle

IGCC Destec (E-Gas^{TM}) / HGCU / "G" Gas Turbine

Destec HGCU IGCC　　　　　　　　　　　　　　　　　　**400 MW POWER PLANT**

Total Plant Investment		PROCESS	PROCESS	1st Q 2002 Dollar COST, K$
AREA NO	PLANT SECTION DESCRIPTION	CONT, %	CONT, K$	W/O CONT
11	Coal Slurry Preparation	0	$0	$25,914
12	Oxygen Plant	0	$0	$46,751
12	Destec Gasifier	5	$2,853	$57,054
12	Gas Compression (Recycle and Quench)	5	$275	$5,491
14	Gas Conditioning	10	$1,532	$15,321
14	Air Boost Compressor	0	$0	$882
14	Transport Desulfurizer	15	$1,322	$8,815
14	Sulfuric Acid Plant	0	$0	$18,554
15	Gas Turbine System	5	$2,868	$57,368
15	HRSG/Steam Turbine	5	$2,454	$49,082
18	Water Systems	0	$0	$9,983
30	Civil/Structural/Architectural	0	$0	$17,684
40	Piping	0	$0	$9,983
50	Control/ Instrumentation	0	$0	$11,409
60	Electrical	0	$0	$24,815

Subtotal, Process Plant Cost				$359,109

Engineering Fees				$35,911
Process Contingency (Using cont. listed)				$11,304
Project Contingency,	15 % Proc Plt & Gen Plt Fac			$53,866

Total Plant Cost (TPC)				$460,191

Plant Construction Period,	4.0 Years (1 or more)			
Construction Interest Rate,	11.2 %			
Adjustment for Interest and Inflation				$57,768

Total Plant Investment (TPI)				$517,958

Prepaid Royalties				$1,796
Initial Catalyst and Chemical Inventory				$302
Startup Costs				$12,548
Spare Parts				$2,301
Working Capital				$5,768
Land,	200 Acres			$1,300

Total Capital Requirement (TCR)				$541,973
			$/kW	1354

ANNUAL OPERATING COSTS

Capacity Factor =   85 %

COST ITEM	QUANTITY	UNIT $ PRICE	ANNUAL COST, K$
Coal (Illinois #6)	2,944 T/D	$24.36 /T	$22,247
Consumable Materials			
Water	2,102 T/D	$0.19 /T	$124
HGCU Sorbent	0.11 T/D	$6,000 /T	$197
Nahcolite	2.3 T/D	$275 /T	$196
Ash/Sorbent Disposal Costs	436 T/D	$8.00 /T	$1,082
Plant Labor			
Oper Labor (incl benef)	15 Men/shift	$34.00 /Hr.	$4,455
Supervision & Clerical			$2,552
Maintenance Costs 2.2%			$10,124
Royalties			$222
Other Operating Costs			$851
Total Operating Costs			**$42,050**
By-Product Credits			
Sulfuric Acid	223.8 T/D	$68.00 /T	$4,722
_____	0.0 T/D	$0.00 /T	$0
_____	0.0 T/D	$0.00 /T	$0
_____	0.0 T/D	$0.00 /T	$0
Total By-Product Credits			$4,722
Net Operating Costs			**$37,328**

B-39

BASES AND ASSUMPTIONS

A. CAPITAL BASES AND DETAILS

	QUANTITY	UNIT $ PRICE	COST, K$
Initial Cat./Chem. Inventory			
Water	53594 T	$0.19 /T	$10
HGCU Sorbent	46 T	$6,000 /T	$276
Nahcolite	59 T	$275 /T	$16
Total Catalyst and Chemical Inventory			$302
Startup costs			
Plant modifications,	2 % TPI		$10,359
Operating costs			$1,651
Fuel			$538
Total Startup Costs			$12,548
Working capital			
Fuel & Consumables inv	60 days supply		$4,402
By-Product inventory	30 days supply		$457
Direct expenses	30 days		$909
Total Working Capital			$5,768

B. ECONOMIC ASSUMPTIONS

Project life	20 Years
Book life	20 Years
Tax life	20 Years
Federal and state income tax rate	38.0 %
Tax depreciation method	MACRS
Investment Tax Credit	0.0 %
Financial structure	

Type of Security	% of Total	Current Dollar Cost, %	Ret, %	Constant Dollar Cost, %	Ret, %
Debt	80	9.0	7.2	5.8	4.6
Preferred Stock	0	3.0	0.0	0.0	0.0
Common Stock	20	20.0	4.0	16.5	3.3
Discount rate (cost of capital)			11.2		7.9

Inflation rate, % per year	3.0
Real Escalation rates (over inflation)	
Fuel, % per year	-1.1
Operating & Maintenance, % per year	0.0

C. COST OF ELECTRICITY

The approach to determining the cost of electricity is based upon the methodology described in the Technical Assessment Guide, published by the Electric Power Research Institute. The cost of electricity is stated in terms of 10th year levelized dollars.

	Current $	Constant $
Levelizing Factors		
Capital Carrying Charge, 10th yr	0.179	0.148
Fuel, 10th year	1.091	0.948
Operating & Maintenance, 10th yr	1.151	1.000

Cost of Electricity - Levelized	mills/kWh	mills/kWh
Capital Charges	32.5	27.0
Fuel Costs	8.1	7.1
Consumables	0.6	0.5
Fixed Operating & Maintenance	6.0	5.2
Variable Operating & Maintenance	1.1	0.9
By-product	-1.8	-1.6
Total Cost of Electricity	46.5	39.1

Combined Cycle

IGCC Destec (E-Gas^{TM}) / CGCU /
"G" Gas Turbine / CO_2 Capture

Destec CGCU IGCC (with CO2 Capture)   **359 MW POWER PLANT**

1st Q 2002 Dollar

		PROCESS	PROCESS	COST, K$
Total Plant Investment		CONT, %	CONT, K$	W/O CONT
AREA NO	PLANT SECTION DESCRIPTION			
11	Coal Slurry Preparation	0	$0	$27,807
12	Oxygen Plant	0	$0	$51,897
12	Destec Gasifier	5	$3,210	$64,202
12	Recycle Gas Compressor	5	$137	$2,750
14	Low Temperature Gas Cooling	0	$0	$15,834
14	Shift Reaction System	0	$0	$16,699
14	SELEXOL (H2S & CO2)	0	$0	$35,125
14	CO2 Compression/Recovery	0	$0	$19,893
14	Claus	0	$0	$9,942
14	SCOT	0	$0	$4,204
15	Gas Turbine System	5	$2,872	$57,448
15	HRSG/Steam Turbine	5	$2,291	$45,814
18	Water Systems	0	$0	$19,339
30	Civil/Structural/Architectural	0	$0	$21,800
40	Piping	0	$0	$19,339
50	Control/ Instrumentation	0	$0	$14,065
60	Electrical	0	$0	$30,590

Subtotal, Process Plant Cost		$456,747

Engineering Fees			$45,675
Process Contingency (Using cont. listed)			$8,511
Project Contingency,	15 % Proc Plt & Gen Plt Fac		$68,512

Total Plant Cost (TPC)	$579,445

Plant Construction Period,	4.0 Years (1 or more)	
Construction Interest Rate,	11.2 %	
Adjustment for Interest and Inflation		$72,737

Total Plant Investment (TPI)	$652,182

Prepaid Royalties		$2,284
Initial Catalyst and Chemical Inventory		$84
Startup Costs		$15,538
Spare Parts		$2,897
Working Capital		$6,062
Land,	200 Acres	$1,300

Total Capital Requirement (TCR)		$680,347
	$/kW	1897

ANNUAL OPERATING COSTS

Capacity Factor =   85 %

COST ITEM	QUANTITY	UNIT $ PRICE	ANNUAL COST, K$
Coal (Illinois #6)	3,256 T/D	$24.36 /T	$24,606
Consumable Materials			
Water	2,924 T/D	$0.19 /T	$172
Selexol Solvent	806.4 Lb/D	$1.45 /Lb	$363
Claus Catalyst	0.01 T/D	$470 /T	$1
SCOT Activated Alum	15.9 Lb/D	$0.67 /Lb	$3
SCOT Cobalt Catalyst			$5
SCOT Chemicals			$16
Ash/Sorbent Disposal Costs	413 T/D	$8.00 /T	$1,024
Plant Labor			
Oper Labor (incl benef)	15 Men/shift	$34.00 /Hr.	$4,455
Supervision & Clerical			$2,866
Maintenance Costs 2.2%			$12,748
Royalties			$246
Other Operating Costs			$955
Total Operating Costs			**$47,461**
By-Product Credits			
Sulfur	79.0 T/D	$75.00 /T	$1,838
_____	0.0 T/D	$0.00 /T	$0
_____	0.0 T/D	$0.00 /T	$0
_____	0.0 T/D	$0.00 /T	$0
Total By-Product Credits			**$1,838**
Net Operating Costs			**$45,624**

BASES AND ASSUMPTIONS

A. CAPITAL BASES AND DETAILS

	QUANTITY	UNIT $ PRICE	COST, K$
Initial Cat./Chem. Inventory			
Water	74550 T	$0.19 /T	$14
Selexol Solvent	20563 Lb	$1.45 /Lb	$30
Claus Catalyst	0.3 T	$470 /T	$0
SCOT Activated Alun	405 Lb	$0.67 /Lb	$0
SCOT Cobalt Catalyst			$16
SCOT Chemicals			$24
Total Catalyst and Chemical Inventory			$84
Startup costs			
Plant modifications,	2 % TPI		$13,044
Operating costs			$1,900
Fuel			$595
Total Startup Costs			$15,538
Working capital			
Fuel & Consumables inv	60 days supply		$4,863
By-Product inventory	30 days supply		$178
Direct expenses	30 days		$1,021
Total Working Capital			$6,062

B. ECONOMIC ASSUMPTIONS

Project life	20 Years
Book life	20 Years
Tax life	20 Years
Federal and state income tax rate	38.0 %
Tax depreciation method	MACRS
Investment Tax Credit	0.0 %
Financial structure	

Type of Security	% of Total	Current Dollar		Constant Dollar	
		Cost, %	Ret, %	Cost, %	Ret, %
Debt	80	9.0	7.2	5.8	4.6
Preferred Stock	0	3.0	0.0	0.0	0.0
Common Stock	20	20.0	4.0	16.5	3.3
Discount rate (cost of capital)			11.2		7.9

Inflation rate, % per year	3.0
Real Escalation rates (over inflation)	
Fuel, % per year	-1.1
Operating & Maintenance, % per year	0.0

C. COST OF ELECTRICITY

The approach to determining the cost of electricity is based upon the
methodology described in the Technical Assessment Guide, published by
the Electric Power Research Institute. The cost of electricity is
stated in terms of 10th year levelized dollars.

	Current $	Constant $
Levelizing Factors		
Capital Carrying Charge, 10th yr	0.179	0.148
Fuel, 10th year	1.091	0.948
Operating & Maintenance, 10th yr	1.151	1.000

Cost of Electricity - Levelized	mills/kWh	mills/kWh
Capital Charges	45.6	37.8
Fuel Costs	10.1	8.7
Consumables	0.7	0.6
Fixed Operating & Maintenance	7.8	6.8
Variable Operating & Maintenance	1.4	1.2
By-product	-0.8	-0.7
Total Cost of Electricity	**64.7**	**54.4**

Combined Cycle

IGCC Shell / CGCU / "G" Gas Turbine

Shell CGCU IGCC Case

413 MW POWER PLANT

1st Q 2002 Dollar

| Total Plant Investment | | PROCESS | PROCESS | COST, K$ |
AREA NO	PLANT SECTION DESCRIPTION	CONT, %	CONT, K$	W/O CONT
11	Coal Preparation	0	$0	$18,436
12	Oxygen Plant	0	$0	$52,564
12	Shell Gasifier	5	$4,041	$80,826
12	Quench Gas Compressor	5	$98	$1,951
14	Low Temperature Gas Cooling/Gas Saturation	0	$0	$9,606
14	MDEA	0	$0	$5,228
14	Claus	0	$0	$10,234
14	SCOT	0	$0	$4,328
15	Gas Turbine System	5	$2,755	$55,107
15	HRSG/Steam Turbine	5	$2,591	$51,828
18	Water Systems	0	$0	$15,956
30	Civil/Structural/Architectural	0	$0	$17,987
40	Piping	0	$0	$15,956
50	Control/ Instrumentation	0	$0	$11,604
60	Electrical	0	$0	$25,239

Subtotal, Process Plant Cost	$376,851

Engineering Fees		$37,685
Process Contingency (Using cont. listed)		$9,486
Project Contingency,	15 % Proc Plt & Gen Plt Fac	$56,528

Total Plant Cost (TPC)	$480,549

Plant Construction Period,	4.0 Years (1 or more)	
Construction Interest Rate,	11.2 %	
Adjustment for Interest and Inflation		$60,323

Total Plant Investment (TPI)	$540,873

Prepaid Royalties		$1,884
Initial Catalyst and Chemical Inventory		$61
Startup Costs		$13,039
Spare Parts		$2,403
Working Capital		$5,788
Land,	200 Acres	$1,300

Total Capital Requirement (TCR)	$565,348
$/kW	1370

ANNUAL OPERATING COSTS

Capacity Factor =   85 %

COST ITEM	QUANTITY	UNIT $ PRICE	ANNUAL COST, K$
Coal (Illinois #6)	3,171 T/D	$24.36 /T	$23,964
Consumable Materials			
Water	1,263 T/D	$0.19 /T	$74
MDEA Solvent	403.2 Lb/D	$1.45 /Lb	$181
Claus Catalyst	0.01 T/D	$470 /T	$1
SCOT Activated Alun	15.9 Lb/D	$0.67 /Lb	$3
SCOT Cobalt Catalyst			$5
SCOT Chemicals			$16
Ash/Sorbent Disposal Costs	321 T/D	$8.00 /T	$797
Plant Labor			
Oper Labor (incl benef)	15 Men/shift	$34.00 /Hr.	$4,455
Supervision & Clerical			$2,605
Maintenance Costs 2.2%			$10,572
Royalties			$240
Other Operating Costs			$868
Total Operating Costs			**$43,783**
By-Product Credits			
Sulfur	78.0 T/D	$75.00 /T	$1,814
_____	0.0 T/D	$0.00 /T	$0
_____	0.0 T/D	$0.00 /T	$0
_____	0.0 T/D	$0.00 /T	$0
Total By-Product Credits			**$1,814**
Net Operating Costs			**$41,969**

BASES AND ASSUMPTIONS

A. CAPITAL BASES AND DETAILS

	QUANTITY	UNIT $ PRICE	COST, K$
Initial Cat./Chem. Inventory			
Water	32212 T	$0.19 /T	$6
MDEA Solvent	10282 Lb	$1.45 /Lb	$15
Claus Catalyst	0.3 T	$470 /T	$0
SCOT Activated Alun	405 Lb	$0.67 /Lb	$0
SCOT Cobalt Catalyst			$16
SCOT Chemicals			$24
Total Catalyst and Chemical Inventory			$61
Startup costs			
Plant modifications,	2 % TPI		$10,817
Operating costs			$1,643
Fuel			$579
Total Startup Costs			$13,039
Working capital			
Fuel & Consumables inv	60 days supply		$4,685
By-Product inventory	30 days supply		$175
Direct expenses	30 days		$928
Total Working Capital			$5,788

B. ECONOMIC ASSUMPTIONS

Project life	20 Years
Book life	20 Years
Tax life	20 Years
Federal and state income tax rate	38.0 %
Tax depreciation method	ACRS
Investment Tax Credit	0.0 %

Financial structure

Type of Security	% of Total	Current Dollar Cost, %	Ret, %	Constant Dollar Cost, %	Ret, %
Debt	80	9.0	7.2	5.8	4.6
Preferred Stock	0	3.0	0.0	0.0	0.0
Common Stock	20	20.0	4.0	16.5	3.3
Discount rate (cost of capital)			11.2		7.9

Inflation rate, % per year	3.0
Real Escalation rates (over inflation)	
Fuel, % per year	-1.1
Operating & Maintenance, % per year	0.0

C. COST OF ELECTRICITY

The approach to determining the cost of electricity is based upon the
methodology described in the Technical Assessment Guide, published by
the Electric Power Research Institute. The cost of electricity is
stated in terms of 10th year levelized dollars.

	Current $	Constant $
Levelizing Factors		
Capital Carrying Charge, 10th yr	0.179	0.148
Fuel, 10th year	1.091	0.948
Operating & Maintenance, 10th yr	1.151	1.000

Cost of Electricity - Levelized	mills/kWh	mills/kWh
Capital Charges	32.9	27.3
Fuel Costs	8.5	7.4
Consumables	0.4	0.4
Fixed Operating & Maintenance	6.0	5.2
Variable Operating & Maintenance	1.1	0.9
By-product	-0.7	-0.6
Total Cost of Electricity	48.2	40.6

Combined Cycle

IGCC Shell / CGCU / "G" Gas Turbine / CO_2 Capture

Shell CGCU I (co2 , h2, power)

351 MW POWER PLANT

Total Plant Investment		PROCESS	PROCESS	1st Q 2002 Dollar COST, K$
AREA NO	PLANT SECTION DESCRIPTION	CONT, %	CONT, K$	W/O CONT
11	Coal Preparation	0	$0	$18,436
12	Oxygen Plant (includes air cpr + O2 cpr)	0	$0	$51,308
12	Shell Gasifier	5	$4,041	$80,826
12	Quench Gas Compressor	5	$98	$1,951
14	Gas Cooling	0	$0	$9,606
14	Shift Reaction System	0	$0	$16,263
14	SELEXOL (H2S & CO2)	0	$0	$29,529
14	CO2 Compression/Recovery	0	$0	$19,374
14	Claus	0	$0	$10,234
14	SCOT	0	$0	$4,328
14	PSA	0	$0	$9,572
15	Gas Turbine System (62 MWe)	5	$627	$12,547
15	HRSG/Steam Turbine	5	$1,215	$24,291
17	Advanced Power System (H2 - fuel cell)	25	$27,518	$110,071
18	Water Systems	0	$0	$21,909
30	Civil/Structural/Architectural	0	$0	$24,697
40	Piping	0	$0	$21,909
50	Control/ Instrumentation	0	$0	$15,933
60	Electrical	0	$0	$34,655

Subtotal, Process Plant Cost		$517,439

Engineering Fees			$51,744
Process Contingency (Using cont. listed)			$33,498
Project Contingency,	15 % Proc Plt & Gen Plt Fac		$77,616

Total Plant Cost (TPC)	$680,297

Plant Construction Period,	4.0 Years (1 or more)	
Construction Interest Rate,	11.2 %	
Adjustment for Interest and Inflation		$85,397

Total Plant Investment (TPI)	$765,695

Prepaid Royalties		$2,587
Initial Catalyst and Chemical Inventory		$76
Startup Costs		$17,972
Spare Parts		$3,401
Working Capital		$6,011
Land,	200 Acres	$1,300

Total Capital Requirement (TCR)		$797,043
	$/kW	2270

ANNUAL OPERATING COSTS

Capacity Factor =   85 %

COST ITEM	QUANTITY	UNIT $ PRICE	ANNUAL COST, K$
Coal (Illinois #6)	3,171 T/D	$24.36 /T	$23,964
Consumable Materials			
Water	1,263 T/D	$0.19 /T	$74
Selexol Solvent	806.4 Lb/D	$1.45 /Lb	$363
Claus Catalyst	0.01 T/D	$470 /T	$1
SCOT Activated Alum	15.9 Lb/D	$0.67 /Lb	$3
SCOT Cobalt Catalyst			$5
SCOT Chemicals			$16
Ash/Sorbent Disposal Costs	321 T/D	$8.00 /T	$797
Plant Labor			
Oper Labor (incl benef)	15 Men/shift	$34.00 /Hr.	$4,455
Supervision & Clerical			$3,133
Maintenance Costs	2.2%		$14,967
Royalties			$240
Other Operating Costs			$1,044
Total Operating Costs			**$49,062**
By-Product Credits			
Sulfur	78.0 T/D	$75.00 /T	$1,814
_____	0.0 T/D	$0.00 /T	$0
_____	0.0 T/D	$0.00 /T	$0
_____	0.0 T/D	$0.00 /T	$0
Total By-Product Credits			$1,814
Net Operating Costs			**$47,248**

BASES AND ASSUMPTIONS

A. CAPITAL BASES AND DETAILS

	QUANTITY	UNIT $ PRICE	COST, K$
Initial Cat./Chem. Inventory			
Water	32212 T	$0.19 /T	$6
Selexol Solvent	20563 Lb	$1.45 /Lb	$30
Claus Catalyst	0.3 T	$470 /T	$0
SCOT Activated Alun	405 Lb	$0.67 /Lb	$0
SCOT Cobalt Catalyst			$16
SCOT Chemicals			$24
Total Catalyst and Chemical Inventory			$76

Startup costs			
Plant modifications,	2 % TPI		$15,314
Operating costs			$2,079
Fuel			$579
Total Startup Costs			$17,972

Working capital			
Fuel & Consumables inv	60 days supply		$4,720
By-Product inventory	30 days supply		$175
Direct expenses	30 days		$1,116
Total Working Capital			$6,011

B. ECONOMIC ASSUMPTIONS

Project life	20 Years
Book life	20 Years
Tax life	20 Years
Federal and state income tax rate	38.0 %
Tax depreciation method	ACRS
Investment Tax Credit	0.0 %
Financial structure	

Type of Security	% of Total	Current Dollar Cost, %	Ret, %	Constant Dollar Cost, %	Ret, %
Debt	80	9.0	7.2	5.8	4.6
Preferred Stock	0	3.0	0.0	0.0	0.0
Common Stock	20	20.0	4.0	16.5	3.3
Discount rate (cost of capital)			11.2		7.9

Inflation rate, % per year	3.0
Real Escalation rates (over inflation)	
Fuel, % per year	-1.1
Operating & Maintenance, % per year	0.0

C. COST OF ELECTRICITY

The approach to determining the cost of electricity is based upon the methodology described in the Technical Assessment Guide, published by the Electric Power Research Institute. The cost of electricity is stated in terms of 10th year levelized dollars.

	Current $	Constant $
Levelizing Factors		
Capital Carrying Charge, 10th yr	0.179	0.148
Fuel, 10th year	1.091	0.948
Operating & Maintenance, 10th yr	1.151	1.000

Cost of Electricity - Levelized	mills/kWh	mills/kWh
Capital Charges	54.6	45.3
Fuel Costs	10.0	8.7
Consumables	0.6	0.5
Fixed Operating & Maintenance	8.9	7.8
Variable Operating & Maintenance	1.6	1.4
By-product	-0.8	-0.7
Total Cost of Electricity	74.8	62.9

Hydraulic Air Compression (HAC)

Natural Gas HAC - No CO_2 Capture

Hydraulic Air Compression Technology Combined Cycle (Natural Gas, No CO2 Capture) Total Plant Investment		324 MW POWER PLANT 1st Q 2002 Dollar		
AREA NO	PLANT SECTION DESCRIPTION	PROCESS CONT, %	PROCESS CONT, K$	COST, K$ W/O CONT
	GT Expander / Combustor	25	$7,980	$31,920
	HV Cpr System	25	$11,736	$46,945
	Well	10	$23	$225
	Recuperator + Air Heaters	0	$0	$6,863
	HRSG/Turbine Package	0	$0	$2,610
18	Water Systems	0	$0	$6,288
30	Civil/Structural/Architectural	0	$0	$12,310
40	Piping	0	$0	$6,288
50	Control/ Instrumentation	0	$0	$7,085
60	Electrical	0	$0	$13,993
	Subtotal, Process Plant Cost		$19,739	$134,527

Engineering Fees		$13,453
Process Contingency (Using cont. listed)		$19,739
Project Contingency,	15 % Proc Plt & Gen Plt Fac	$20,179

Total Plant Cost (TPC)	$187,897

Plant Construction Period,	2.0 Years (1 or more)	
Construction Interest Rate,	11.2 %	
Adjustment for Interest and Inflation		$7,460

Total Plant Investment (TPI)	$195,357

Prepaid Royalties		$673
Initial Catalyst and Chemical Inventory		$68
Startup Costs		$10,822
Spare Parts		$939
Working Capital		$12,417
Land,	100 Acres @ $1500/acre	$150

Total Capital Requirement (TCR)	$220,425
$/kW	681

ANNUAL OPERATING COSTS

Capacity Factor = 85 %

Consumables COST ITEM	QUANTITY		UNIT $ UNIT $ PRICE		ANNUAL ANNUAL COST, K$
Natural Gas	54,568	1000 SCF/D	$3.20	$/1000 SCF	$54,175
Water	562,826	T/d	$0.05	/T	$8,294
Plant Labor					
Oper Labor (incl benef)	5	Men/shift	$34.00	/Hr.	$1,485
Supervision & Clerical					$942
Maintenance Costs	2.2%				$4,134
Insurance & Local Taxes					$3,758
Other Operating Costs					$314
Total Operating Costs					**$73,102**

BASES AND ASSUMPTIONS

A. CAPITAL BASES AND DETAILS

	QUANTITY	UNIT $ PRICE	COST, K$
Initial Cat./Chem. Inventory			
Water	1,435,206 tons	$0.05 /T	$68
Total Catalyst and Chemical Inventory			$68
Startup costs			
Plant modifications,	2 % TPI		$3,907
Operating costs			$6,914
Total Startup Costs			$10,822
Working capital			
Fuel & Consumables inv	60 days supply		$12,081
Direct expenses	30 days		$335
Total Working Capital			$12,417

B. ECONOMIC ASSUMPTIONS

Project life	20 Years
Book life	20 Years
Tax life	20 Years
Federal and state income tax rate	38.0 %
Tax depreciation method	MACRS
Investment Tax Credit	0.0 %

Financial structure

Type of Security	% of Total	Current Dollar Cost, %	Current Dollar Ret, %	Constant Dollar Cost, %	Constant Dollar Ret, %
Debt	80	9.0	7.2	5.8	4.6
Preferred Stock	0	0.2	0.0	0.0	0.0
Common Stock	20	20.0	4.0	16.5	3.3
Discount rate (cost of capital)		11.2			7.9

Inflation rate, % per year	3.0
Real Escalation rates (over inflation)	
Fuel, % per year	0.2
Operating & Maintenance, % per year	0.0

C. COST OF ELECTRICITY

The approach to determining the cost of electricity is based upon the methodology described in the Technical Assessment Guide, published by the Electric Power Research Institute. The cost of electricity is stated in terms of 10th year levelized dollars.

	Current $	Constant $
Levelizing Factors		
Capital Carrying Charge, 10th yr	0.179	0.148
Fuel, 10th year	1.162	1.010
Operating & Maintenance, 10th yr	1.151	1.000

Cost of Electricity - Levelized	mills/kWh	mills/kWh
Capital Charges	16.4	13.6
Fuel Costs	26.1	22.7
Consumables	4.0	3.4
Fixed Operating & Maintenance	4.3	3.8
Variable Operating & Maintenance	0.8	0.7
By-product	0.0	0.0
Total Cost of Electricity	51.6	44.2

Hydraulic Air Compression (HAC)

Natural Gas HAC - CO_2 Capture

		PROCESS	PROCESS	COST, K$
Hydraulic Air Compression Technology Combined Cycle			**300 MW POWER PLANT**	
(Natural Gas, CO2 Capture)				1st Q 2002 Dollar
Total Plant Investment		PROCESS	PROCESS	COST, K$
AREA NO	PLANT SECTION DESCRIPTION	CONT, %	CONT, K$	W/O CONT
	GT Expander / Combustor	25	$7,980	$31,920
	HV Cpr System	25	$11,736	$46,945
	Well	10	$23	$225
	Recuperator + Air Heaters	0	$0	$8,215
	HRSG	0	$0	$5,837
	Amine System	5	$3,541	$70,815
	CO2 Compression/drying	5	$574	$11,488
18	Water Systems	0	$0	$6,613
30	Civil/Structural/Architectural	0	$0	$12,947
40	Piping	0	$0	$6,613
50	Control/ Instrumentation	0	$0	$7,451
60	Electrical	0	$0	$14,716
	Subtotal, Process Plant Cost		$23,854	$223,786

Engineering Fees			$22,379
Process Contingency (Using cont. listed)			$19,739
Project Contingency,	15 % Proc Plt & Gen Plt Fac		$33,568
	Total Plant Cost (TPC)		$299,471

Plant Construction Period,	2.0 Years (1 or more)		
Construction Interest Rate,	11.2 %		
Adjustment for Interest and Inflation			$11,889
	Total Plant Investment (TPI)		$311,360

Prepaid Royalties			$1,119
Initial Catalyst and Chemical Inventory			$68
Startup Costs			$14,219
Spare Parts			$1,497
Working Capital			$13,841
Land,	100 Acres	@ $1500/acre	$150
	Total Capital Requirement (TCR)		$342,254
		$/kW	1140

ANNUAL OPERATING COSTS

Capacity Factor =   85 %

Consumables			UNIT $		ANNUAL
COST ITEM	QUANTITY		PRICE		COST, K$
Natural Gas	61,439 1000 SCF/D		$3.20	$/1000 SCF	$60,997
Water	562,826 T/d		$0.05	/T	$8,294
Amine Chemicals	141 ton CO2/hr		$3.00	/ton CO2 Captu	$3,147
Plant Labor					
Oper Labor (incl benef)	5 Men/shift		$34.00	/Hr.	$1,485
Supervision & Clerical					$1,236
Maintenance Costs	2.2%				$6,588
Insurance & Local Taxes					$5,989
Other Operating Costs					$412
Total Operating Costs					$85,002

BASES AND ASSUMPTIONS

A. CAPITAL BASES AND DETAILS

	QUANTITY	UNIT $ PRICE	COST, K$
Initial Cat./Chem. Inventory			
Water	1,435,206 tons	$0.05 /T	$68
Amine Chemicals	20,120 (7 days CO2)	$3.00 /ton CO2	$60
Total Catalyst and Chemical Inventory			$68
Startup costs			
Plant modifications,	2 % TPI		$6,227
Operating costs			$7,992
Total Startup Costs			$14,219
Working capital			
Fuel & Consumables inv	60 days supply		$13,400
Direct expenses	30 days		$440
Total Working Capital			$13,841

B. ECONOMIC ASSUMPTIONS

Project life	20 Years
Book life	20 Years
Tax life	20 Years
Federal and state income tax rate	38.0 %
Tax depreciation method	MACRS
Investment Tax Credit	0.0 %

Financial structure

Type of Security	% of Total	Current Dollar Cost, %	Current Dollar Ret, %	Constant Dollar Cost, %	Constant Dollar Ret, %
Debt	80	9.0	7.2	5.8	4.6
Preferred Stock	0	0.2	0.0	0.0	0.0
Common Stock	20	20.0	4.0	16.5	3.3
Discount rate (cost of capital)			11.2		7.9

Inflation rate, % per year	3.0
Real Escalation rates (over inflation)	
Fuel, % per year	0.2
Operating & Maintenance, % per year	0.0

C. COST OF ELECTRICITY

The approach to determining the cost of electricity is based upon the methodology described in the Technical Assessment Guide, published by the Electric Power Research Institute. The cost of electricity is stated in terms of 10th year levelized dollars.

	Current $	Constant $
Levelizing Factors		
Capital Carrying Charge, 10th yr	0.179	0.148
Fuel, 10th year	1.162	1.010
Operating & Maintenance, 10th yr	1.151	1.000

Cost of Electricity - Levelized	mills/kWh	mills/kWh
Capital Charges	27.4	22.7
Fuel Costs	31.7	27.6
Consumables	4.3	3.7
Fixed Operating & Maintenance	6.9	6.0
Variable Operating & Maintenance	1.2	1.1
By-product	0.0	0.0
Total Cost of Electricity	71.5	61.0

Hydraulic Air Compression (HAC)

Coal Syngas HAC
    Destec (E-Gas^{TM}) / CGCU / "G" GT / No CO_2 Capture

Destec Gasification / CGCU / HAC
Hydraulic Air Compression Technology Combined Cycle   **326 MW POWER PLANT**
           (COAL, No CO2 Capture)   1st Q 2002 Dollar

Total Plant Investment		PROCESS	PROCESS	COST, K$
AREA NO	PLANT SECTION DESCRIPTION	CONT, %	CONT, K$	W/O CONT
11	Coal Slurry Preparation	0	$0	$24,429
12	Oxygen Plant	0	$0	$38,848
12	Destec Gasifier	5	$1,008	$20,163
12	Recycle Gas Compressor / Fuel Coolers	5	$124	$2,484
14	Low Temperature Gas Cooling	0	$0	$13,824
14	MDEA	0	$0	$4,894
14	Claus	0	$0	$9,218
14	SCOT	0	$0	$3,898
15	Hydraulic Air Compression System	25	$12,371	$49,485
15	GT Expander / Combustor	25	$7,980	$31,920
15	Well	10	$23	$225
15	Recuperator	0	$0	$16,000
15	HRSG/Steam Turbine	5	$761	$15,221
18	Water Systems	0	$0	$12,684
30	Civil/Structural/Architectural	0	$0	$14,298
40	Piping	0	$0	$12,684
50	Control/ Instrumentation	0	$0	$9,224
60	Electrical	0	$0	$20,063

Subtotal, Process Plant Cost	$299,565

Engineering Fees		$29,956
Process Contingency (Using cont. listed)		$22,267
Project Contingency,	15 % Proc Plt & Gen Plt Fac	$44,935

Total Plant Cost (TPC)	$396,723

Plant Construction Period,	4.0 Years (1 or more)	
Construction Interest Rate,	11.2 %	
Adjustment for Interest and Inflation		$49,800

Total Plant Investment (TPI)	$446,523

Prepaid Royalties		$1,498
Initial Catalyst and Chemical Inventory		$806
Startup Costs		$10,908
Spare Parts		$1,984
Working Capital		$5,022
Land,	200 Acres	$1,300

Total Capital Requirement (TCR)	$468,041
$/kW	1436

ANNUAL OPERATING COSTS

Capacity Factor = 85 %

COST ITEM	QUANTITY	UNIT $ PRICE	ANNUAL COST, K$
Coal (Illinois #6)	2,706 T/D	$24.36 /T	$20,449
Consumable Materials			
Water	3,073 T/D	$0.19 /T	$181
HAC Makeup Water	607,513 T/D	$0.05 /T	$8,953
MDEA Solvent	403.2 Lb/D	$1.45 /Lb	$181
Claus Catalyst	0.01 T/D	$470 /T	$1
SCOT Activated Alum	15.9 Lb/D	$0.67 /Lb	$3
SCOT Cobalt Catalyst			$5
SCOT Chemicals			$16
Ash/Sorbent Disposal Costs	358 T/D	$8.00 /T	$889
Plant Labor			
Oper Labor (incl benef)	15 Men/shift	$34.00 /Hr.	$4,455
Supervision & Clerical			$2,384
Maintenance Costs 2.2%			$8,728
Royalties			$204
Other Operating Costs			$795
Total Operating Costs			**$47,246**
By-Product Credits			
Sulfur	65.4 T/D	$75.00 /T	$1,521
_____	0.0 T/D	$0.00 /T	$0
_____	0.0 T/D	$0.00 /T	$0
_____	0.0 T/D	$0.00 /T	$0
Total By-Product Credits			$1,521
Net Operating Costs			**$45,725**

BASES AND ASSUMPTIONS

A. CAPITAL BASES AND DETAILS

	QUANTITY	UNIT $ PRICE		COST, K$
Initial Cat./Chem. Inventory				
Water	78368 T	$0.19	/T	$15
HAC Makeup Water	15491570 T	$0.05	/T	$736
MDEA Solvent	10282 Lb	$1.45	/Lb	$15
Claus Catalyst	0.3 T	$470	/T	$0
SCOT Activated Alun	405 Lb	$0.67	/Lb	$0
SCOT Cobalt Catalyst				$16
SCOT Chemicals				$24
Total Catalyst and Chemical Inventory				$806
Startup costs				
Plant modifications,	2 % TPI			$8,930
Operating costs				$1,483
Fuel				$494
Total Startup Costs				$10,908
Working capital				
Fuel & Consumables inv	60 days supply			$4,026
By-Product inventory	30 days supply			$147
Direct expenses	30 days			$849
Total Working Capital				$5,022

B. ECONOMIC ASSUMPTIONS

Project life	20 Years
Book life	20 Years
Tax life	20 Years
Federal and state income tax rate	38.0 %
Tax depreciation method	MACRS
Investment Tax Credit	0.0 %
Financial structure	

Type of Security	% of Total	Current Dollar Cost, %	Ret, %	Constant Dollar Cost, %	Ret, %
Debt	80	9.0	7.2	5.8	4.6
Preferred Stock	0	3.0	0.0	0.0	0.0
Common Stock	20	20.0	4.0	16.5	3.3
Discount rate (cost of capital)			11.2		7.9

Inflation rate, % per year	3.0
Real Escalation rates (over inflation)	
Fuel, % per year	-1.1
Operating & Maintenance, % per year	0.0

C. COST OF ELECTRICITY

The approach to determining the cost of electricity is based upon the
methodology described in the Technical Assessment Guide, published by
the Electric Power Research Institute. The cost of electricity is
stated in terms of 10th year levelized dollars.

	Current $	Constant $
Levelizing Factors		
Capital Carrying Charge, 10th yr	0.179	0.148
Fuel, 10th year	1.091	0.948
Operating & Maintenance, 10th yr	1.151	1.000

Cost of Electricity - Levelized	mills/kWh	mills/kWh
Capital Charges	34.5	28.6
Fuel Costs	9.2	8.0
Consumables	4.9	4.2
Fixed Operating & Maintenance	6.7	5.8
Variable Operating & Maintenance	1.2	1.0
By-product	-0.7	-0.6
Total Cost of Electricity	55.7	47.0

Hydraulic Air Compression (HAC)

Coal Syngas HAC
Destec High Pressure (E-Gas^{TM}) / HGCU /
    "G" GT / CO_2 Capture

Hydraulic Air Compression Technology Combined Cycle
Destec Gasification / HGCU / HSD   **312 MW POWER PLANT**
                     (COAL, CO2 Capture)   1st Q 2002 Dollar

Total Plant Investment

AREA NO	PLANT SECTION DESCRIPTION	PROCESS CONT, %	PROCESS CONT, K$	COST, K$ W/O CONT
11	Coal Slurry Preparation	0	$0	$27,689
12	Oxygen Plant	0	$0	$41,336
12	Destec Gasifier/Syngas Cooler	5	$4,170	$83,394
12	Recycle Compressors	5	$40	$790
14	Gas Conditioning	10	$714	$7,139
14	Transport Desulfurizer	15	$847	$5,646
14	Sulfuric Acid Plant	0	$0	$19,930
14	Hydrogen Separation Device	50	$5,041	$10,081
15	CO2 Compressor	0	$0	$28,491
15	H2 Compressor	0	$0	$4,609
15	Gas Expander	0	$0	$6,844
15	Hydraulic Air Compression System	25	$5,315	$53,161
15	GT Expander / Combustor	25	$7,980	$31,920
15	Well	10	$195	$225
15	Recuperator	0	$0	$15,500
15	HRSG/Steam Turbine	5	$818	$16,361
18	Water Systems	0	$0	$12,359
30	Civil/Structural/Architectural	0	$0	$21,893
40	Piping	0	$0	$12,359
50	Control/ Instrumentation	0	$0	$14,125
60	Electrical	0	$0	$30,721

Subtotal, Process Plant Cost	$444,571

Engineering Fees	$44,457
Process Contingency (Using cont. listed)	$25,119
Project Contingency, 15 % Proc Plt & Gen Plt Fac	$66,686

Total Plant Cost (TPC)	$580,833

Plant Construction Period, 4.0 Years (1 or more)	
Construction Interest Rate, 11.2 %	
Adjustment for Interest and Inflation	$72,912

Total Plant Investment (TPI)	$653,745

Prepaid Royalties	$2,223
Initial Catalyst and Chemical Inventory	$1,710
Startup Costs	$15,587
Spare Parts	$2,904
Working Capital	$6,366
Land, 200 Acres	$1,300

Total Capital Requirement (TCR)	$683,834
$/kW	2189

ANNUAL OPERATING COSTS

Capacity Factor =   85 %

COST ITEM	QUANTITY	UNIT $ PRICE	ANNUAL COST, K$
Coal (Illinois #6)	3,236 T/D	$24.36 /T	$24,456
Consumable Materials			
Process Water	4,820 T/D	$0.19 /T	$284
HAC Makeup Water	673,200 T/D	$0.05 /T	$9,921
HGCU Sorbent	0.03 T/D	$6,000 /T	$65
Nahcolite	2.3 T/D	$275 /T	$196
Ash/Sorbent Disposal Costs	487 T/D	$8.00 /T	$1,209
Plant Labor			
Oper Labor (incl benef)	15 Men/shift	$34.00 /Hr.	$4,455
Supervision & Clerical			$2,870
Maintenance Costs 2.2%			$12,778
Royalties			$245
Other Operating Costs			$957

Total Operating Costs			$57,435

By-Product Credits			
Sulfuric Acid	249.4 T/D	$68.00 /T	$5,262
_____	0.0 T/D	$0.00 /T	$0
_____	0.0 T/D	$0.00 /T	$0
_____	0.0 T/D	$0.00 /T	$0
Total By-Product Credits			$5,262

Net Operating Costs			$52,173

BASES AND ASSUMPTIONS

A. CAPITAL BASES AND DETAILS

	QUANTITY	UNIT $ PRICE	COST, K$
Initial Cat./Chem. Inventory			
Water	122909 T	$0.19 /T	$23
HAC Makeup Water	17166600 T	$0.05 /T	$815
HGCU Sorbent	15 T	$6,000 /T	$16
Nahcolite	59 T	$275 /T	$855
Total Catalyst and Chemical Inventory			$1,710
Startup costs			
Plant modifications,	2 % TPI		$13,075
Operating costs			$1,921
Fuel			$591
Total Startup Costs			$15,587
Working capital			
Fuel & Consumables inv	60 days supply		$4,835
By-Product inventory	30 days supply		$509
Direct expenses	30 days		$1,022
Total Working Capital			$6,366

B. ECONOMIC ASSUMPTIONS

Project life	20 Years
Book life	20 Years
Tax life	20 Years
Federal and state income tax rate	38.0 %
Tax depreciation method	MACRS
Investment Tax Credit	0.0 %

Financial structure

Type of Security	% of Total	Current Dollar Cost, %	Ret, %	Constant Dollar Cost, %	Ret, %
Debt	80	9.0	7.2	5.8	4.6
Preferred Stock	0	3.0	0.0	0.0	0.0
Common Stock	20	20.0	4.0	16.5	3.3
Discount rate (cost of capital)			11.2		7.9

Inflation rate, % per year	3.0
Real Escalation rates (over inflation)	
Fuel, % per year	-1.1
Operating & Maintenance, % per year	0.0

C. COST OF ELECTRICITY

The approach to determining the cost of electricity is based upon the methodology described in the Technical Assessment Guide, published by the Electric Power Research Institute. The cost of electricity is stated in terms of 10th year levelized dollars.

	Current $	Constant $
Levelizing Factors		
Capital Carrying Charge, 10th yr	0.179	0.148
Fuel, 10th year	1.091	0.948
Operating & Maintenance, 10th yr	1.151	1.000

Cost of Electricity - Levelized	mills/kWh	mills/kWh
Capital Charges	52.6	43.6
Fuel Costs	11.5	10.0
Consumables	5.8	5.0
Fixed Operating & Maintenance	9.0	7.8
Variable Operating & Maintenance	1.6	1.4
By-product	-2.6	-2.3
Total Cost of Electricity	77.8	65.5

Rocket Engine (CES) - CO_2 Capture

Natural Gas CES (gas generator)

Natural Gas CES   **398 MW POWER PLANT**

| Total Plant Investment | | PROCESS | PROCESS | 1st Q 2002 Dollar COST, K$ |
AREA NO	PLANT SECTION DESCRIPTION	CONT, %	CONT, K$	W/O CONT
12	Oxygen Plant	0	$0	$117,982
14	CH4 Compressor	0	$0	$796
14	Gas Generator + Reheater	25	$1,615	$6,460
15	CO2 Compressor	10	$3,151	$31,513
15	CES Turbines	25	$5,216	$20,864
18	Water Systems	0	$0	$12,611
30	Civil/Structural/Architectural	0	$0	$24,688
40	Piping	0	$0	$12,611
50	Control/ Instrumentation	0	$0	$14,209
60	Electrical	0	$0	$28,063

Subtotal, Process Plant Cost	$269,796

Engineering Fees		$26,980
Process Contingency (Using cont. listed)		$9,982
Project Contingency,	15 % Proc Plt & Gen Plt Fac	$40,469

Total Plant Cost (TPC)	$347,228

Plant Construction Period,	3.0 Years (1 or more)	
Construction Interest Rate,	11.2 %	
Adjustment for Interest and Inflation		$28,300

Total Plant Investment (TPI)	$375,527

Prepaid Royalties			$1,349
Initial Catalyst and Chemical Inventory			$16
Startup Costs			$8,662
Spare Parts			$1,736
Working Capital			$697
Land,	200 Acres	@ $1500/acre	$300

Total Capital Requirement (TCR)	$388,288
$/kW	975

ANNUAL OPERATING COSTS

Capacity Factor = 85 %

COST ITEM	QUANTITY	UNIT $ PRICE	ANNUAL COST, K$
Natural Gas	74,066 1000 SCF/day	$3.20 /1000 SCF	$73,533
Consumable Materials			
Water	3,388 T/D	$0.19 /T	$200
Ash/Sorbent Disposal Costs	0 T/D	$8.00 /T	$0
Plant Labor			
Oper Labor (incl benef)	10 Men/shift	$34.00 /Hr.	$2,970
Supervision & Clerical			$1,808
Maintenance Costs	2.2%		$7,639
Royalties			$735
Other Operating Costs			$603
Total Operating Costs			**$87,488**
By-Product Credits			
	0.0 T/D	$0.00 /T	$0
_____	0.0 T/D	$0.00 /T	$0
_____	0.0 T/D	$0.00 /T	$0
_____	0.0 T/D	$0.00 /T	$0
Total By-Product Credits			$0
Net Operating Costs			**$87,488**

BASES AND ASSUMPTIONS

A. CAPITAL BASES AND DETAILS

	QUANTITY	UNIT $ PRICE	COST, K$
Initial Cat./Chem. Inventory			
Water	86387 T	$0.19 /T	$16
Total Catalyst and Chemical Inventory			$16
Startup costs			
Plant modifications,	2 % TPI		$7,511
Operating costs			1,149.85
Fuel			$2
Total Startup Costs			$8,662
Working capital			
Fuel & Consumables inv	60 days supply		$53
By-Product inventory	30 days supply		$0
Direct expenses	30 days		$644
Total Working Capital			$697

B. ECONOMIC ASSUMPTIONS

Project life	20 Years
Book life	20 Years
Tax life	20 Years
Federal and state income tax rate	38.0 %
Tax depreciation method	MACRS
Investment Tax Credit	0.0 %
Financial structure	

Type of Security	% of Total	Current Dollar Cost, %	Ret, %	Constant Dollar Cost, %	Ret, %
Debt	80	9.0	7.2	5.8	4.6
Preferred Stock	0	3.0	0.0	0.0	0.0
Common Stock	20	20.0	4.0	16.5	3.3
Discount rate (cost of capital)			11.2		7.9

Inflation rate, % per year	3.0
Real Escalation rates (over inflation)	
Fuel, % per year	0.2
Operating & Maintenance, % per year	0.0

C. COST OF ELECTRICITY

The approach to determining the cost of electricity is based upon the
methodology described in the Technical Assessment Guide, published by
the Electric Power Research Institute. The cost of electricity is
stated in terms of 10th year levelized dollars.

	Current $	Constant $
Levelizing Factors		
Capital Carrying Charge, 10th yr	0.179	0.148
Fuel, 10th year	1.162	1.010
Operating & Maintenance, 10th yr	1.151	1.000

Cost of Electricity - Levelized	mills/kWh	mills/kWh
Capital Charges	23.4	19.4
Fuel Costs	28.8	25.0
Consumables	0.1	0.1
Fixed Operating & Maintenance	4.5	3.9
Variable Operating & Maintenance	0.8	0.7
By-product	0.0	0.0
Total Cost of Electricity	57.7	49.2

Rocket Engine (CES) - CO_2 Capture

Coal Syngas CES (gas generator) – Destec HP / HGCU

Destec Coal CES   **406 MW POWER PLANT**

Total Plant Investment AREA NO	PLANT SECTION DESCRIPTION	PROCESS CONT, %	PROCESS CONT, K$	1st Q 2002 Dollar COST, K$ W/O CONT
11	Coal Slurry Preparation	0	$0	$29,661
12	Oxygen Plant	0	$0	$132,368
12	Destec Gasifier	5	$4,259	$85,172
14	Gas Conditioning	10	$912	$9,118
14	Transport Desulfurizer	15	$882	$5,879
14	Sulfuric Acid Plant	0	$0	$21,301
14	Gas Generator + Reheator	25	$1,646	$6,584
15	CO2 Compressor	10	$6,016	$60,164
15	CES Turbines	25	$6,510	$26,039
18	Water Systems	0	$0	$13,170
30	Civil/Structural/Architectural	0	$0	$23,330
40	Piping	0	$0	$13,170
50	Control/ Instrumentation	0	$0	$15,051
60	Electrical	0	$0	$32,737

Subtotal, Process Plant Cost		$473,742

Engineering Fees			$47,374
Process Contingency (Using cont. listed)			$20,224
Project Contingency,	15 % Proc Plt & Gen Plt Fac		$71,061

Total Plant Cost (TPC)	$612,402

Plant Construction Period,	4.0 Years (1 or more)	
Construction Interest Rate,	11.2 %	
Adjustment for Interest and Inflation		$76,875

Total Plant Investment (TPI)	$689,276

Prepaid Royalties		$2,369
Initial Catalyst and Chemical Inventory		$122
Startup Costs		$16,319
Spare Parts		$3,062
Working Capital		$6,898
Land,	200 Acres @ $1500/acre	$300

Total Capital Requirement (TCR)		$718,346
	$/kW	1768

ANNUAL OPERATING COSTS

Capacity Factor = 85 %

COST ITEM	QUANTITY	UNIT $ PRICE	ANNUAL COST, K$
Coal (Illinois #6)	3,570 T/D	$24.36 /T	$26,982
Consumable Materials			
Water	1,187 T/D	$0.19 /T	$70
HGCU Sorbent	0.04 T/D	$6,000 /T	$72
Nahcolite	2.3 T/D	$275 /T	$196
Ash/Sorbent Disposal Costs	121 T/D	$8.00 /T	$299
Plant Labor			
Oper Labor (incl benef)	15 Men/shift	$34.00 /Hr.	$4,455
Supervision & Clerical			$2,953
Maintenance Costs 2.2%			$13,473
Royalties			$270
Other Operating Costs			$984
Total Operating Costs			**$49,754**
By-Product Credits			
Sulfuric Acid	275.9 T/D	$68.00 /T	$5,820
_____	0.0 T/D	$0.00 /T	$0
_____	0.0 T/D	$0.00 /T	$0
_____	0.0 T/D	$0.00 /T	$0
Total By-Product Credits			$5,820
Net Operating Costs			**$43,934**

BASES AND ASSUMPTIONS

A. CAPITAL BASES AND DETAILS

	QUANTITY	UNIT $ PRICE	COST, K$
Initial Cat./Chem. Inventory			
Water	30277 T	$0.19 /T	$6
HGCU Sorbent	17 T	$6,000 /T	$100
Nahcolite	59 T	$275 /T	$16
Total Catalyst and Chemical Inventory			$122
Startup costs			
Plant modifications,	2 % TPI		$13,786
Operating costs			$1,881
Fuel			$652
Total Startup Costs			$16,319
Working capital			
Fuel & Consumables inv	60 days supply		$5,283
By-Product inventory	30 days supply		$563
Direct expenses	30 days		$1,052
Total Working Capital			$6,898

B. ECONOMIC ASSUMPTIONS

Project life	20 Years
Book life	20 Years
Tax life	20 Years
Federal and state income tax rate	38.0 %
Tax depreciation method	MACRS
Investment Tax Credit	0.0 %
Financial structure	

Type of Security	% of Total	Current Dollar Cost, %	Ret, %	Constant Dollar Cost, %	Ret, %
Debt	80	9.0	7.2	5.8	4.6
Preferred Stock	0	3.0	0.0	0.0	0.0
Common Stock	20	20.0	4.0	16.5	3.3
Discount rate (cost of capital)			11.2		7.9

Inflation rate, % per year	3.0
Real Escalation rates (over inflation)	
Fuel, % per year	-1.1
Operating & Maintenance, % per year	0.0

C. COST OF ELECTRICITY

The approach to determining the cost of electricity is based upon the
methodology described in the Technical Assessment Guide, published by
the Electric Power Research Institute. The cost of electricity is
stated in terms of 10th year levelized dollars.

	Current $	Constant $
Levelizing Factors		
Capital Carrying Charge, 10th yr	0.179	0.148
Fuel, 10th year	1.091	0.948
Operating & Maintenance, 10th yr	1.151	1.000

Cost of Electricity - Levelized	mills/kWh	mills/kWh
Capital Charges	42.5	35.3
Fuel Costs	9.7	8.5
Consumables	0.2	0.2
Fixed Operating & Maintenance	7.2	6.2
Variable Operating & Maintenance	1.3	1.1
By-product	-2.2	-1.9
Total Cost of Electricity	58.7	49.3

Hydrogen Turbine - CO$_2$ Capture

Hydrogen from Steam Methane Reforming (SMR)

Hydrogen Turbine Cycle - NATURAL GAS   **413 MW POWER PLANT**

Total Plant Investment

AREA NO	PLANT SECTION DESCRIPTION	PROCESS CONT, %	PROCESS CONT, K$	1st Q 2002 Dollar COST, K$ W/O CONT
	Gas Turbine	5	$2,649	$52,986
	Steam Cycle	5	$2,436	$48,721
	Hydrogen Production	5	$8,375	$167,505
	CO2 Compressor	0	$0	$13,605
18	Water Systems	0	$0	$12,859
30	Civil/Structural/Architectural	0	$0	$25,174
40	Piping	0	$0	$12,859
50	Control/ Instrumentation	0	$0	$14,489
60	Electrical	0	$0	$28,615

Subtotal, Process Plant Cost	$376,813

Engineering Fees		$37,681
Process Contingency (Using cont. listed)		$13,461
Project Contingency,	15 % Proc Plt & Gen Plt Fac	$56,522

Total Plant Cost (TPC)	$484,476

Plant Construction Period,	2.0 Years (1 or more)	
Construction Interest Rate,	11.2 %	
Adjustment for Interest and Inflation		$19,234

Total Plant Investment (TPI)	$503,710

Prepaid Royalties		$1,884
Initial Catalyst and Chemical Inventory		$4
Startup Costs		$20,643
Spare Parts		$2,422
Working Capital		$17,602
Land,	100 Acres @ $1500/acre	$150

Total Capital Requirement (TCR)	$546,415
$/kW	1323

ANNUAL OPERATING COSTS

Capacity Factor =   85 %

Consumables COST ITEM		QUANTITY	UNIT $ UNIT $ PRICE	ANNUAL ANNUAL COST, K$
	Natural Gas	86,047 MMBtu/D	$3.24 $/MMBtu	$86,538
	Water	8,175 T/d	$0.19 /T	$482
Plant Labor				
Oper Labor (incl benef)		10 Men/shift	$34.00 /Hr.	$2,970
Supervision & Clerical				$2,170
Maintenance Costs	2.2%			$10,658
Insurance & Local Taxes				$9,690
Other Operating Costs				$723
Total Operating Costs				$113,232

BASES AND ASSUMPTIONS

A. CAPITAL BASES AND DETAILS

	QUANTITY	UNIT $ PRICE	COST, K$
Initial Cat./Chem. Inventory			
Water	20,847 tons	$0.19 /T	$4
Total Catalyst and Chemical Inventory			$4
Startup costs			
Plant modifications,	2 % TPI		$10,074
Operating costs			$10,569
Total Startup Costs			$20,643
Working capital			
Fuel & Consumables inv	60 days supply		$16,829
Direct expenses	30 days		$773
Total Working Capital			$17,602

B. ECONOMIC ASSUMPTIONS

Project life	20 Years
Book life	20 Years
Tax life	20 Years
Federal and state income tax rate	38.0 %
Tax depreciation method	MACRS
Investment Tax Credit	0.0 %

Financial structure

Type of Security	% of Total	Current Dollar Cost, %	Ret, %	Constant Dollar Cost, %	Ret, %
Debt	80	9.0	7.2	5.8	4.6
Preferred Stock	0	0.2	0.0	0.0	0.0
Common Stock	20	20.0	4.0	16.5	3.3
Discount rate (cost of capital)			11.2		7.9

Inflation rate, % per year	3.0
Real Escalation rates (over inflation)	
Fuel, % per year	0.2
Operating & Maintenance, % per year	0.0

C. COST OF ELECTRICITY

The approach to determining the cost of electricity is based upon the methodology
described in the Technical Assessment Guide, published by the Electric Power
Research Institute. The cost of electricity is stated in terms of 10th year
levelized dollars.

	Current $	Constant $
Levelizing Factors		
Capital Carrying Charge, 10th yr	0.179	0.148
Fuel, 10th year	1.162	1.010
Operating & Maintenance, 10th yr	1.151	1.000

Cost of Electricity - Levelized	mills/kWh	mills/kWh
Capital Charges	31.8	26.4
Fuel Costs	32.7	28.4
Consumables	0.2	0.2
Fixed Operating & Maintenance	8.3	7.2
Variable Operating & Maintenance	1.5	1.3
By-product	0.0	0.0
Total Cost of Electricity	74.5	63.5

Hydrogen Turbine - CO$_2$ Capture

Destec High Pressure (E-Gas^{TM}) / HGCU / HSD

H2 TURBINE COAL (DESTEC)   **376 MW POWER PLANT**

1st Q 2002 Dollar

Total Plant Investment

AREA NO	PLANT SECTION DESCRIPTION	PROCESS CONT, %	PROCESS CONT, K$	COST, K$ W/O CONT
11	Coal Slurry Preparation	0	$0	$29,661
12	Oxygen Plant	0	$0	$62,455
12	Destec Gasifier	5	$3,447	$68,947
14	Gas Conditioning	10	$765	$7,649
14	Transport Desulfurizer	15	$881	$5,871
14	Sulfuric Acid Plant	0	$0	$21,301
14	Hydrogen Separation Device	50	$5,407	$10,814
15	CO2 Compressor	0	$0	$31,670
15	H2 Compressor	0	$0	$6,478
15	Power Turbine	0	$0	$10,339
15	Gas Turbine + Steam Cycle System	5	$4,639	$92,785
15	HRSG/Steam Turbine	0	$0	$19,067
18	Water Systems	0	$0	$20,187
30	Civil/Structural/Architectural	0	$0	$22,756
40	Piping	0	$0	$20,187
50	Control/ Instrumentation	0	$0	$14,681
60	Electrical	0	$0	$31,932

Subtotal, Process Plant Cost	$476,781

Engineering Fees		$47,678
Process Contingency (Using cont. listed)		$15,139
Project Contingency,	15 % Proc Plt & Gen Plt Fac	$71,517

Total Plant Cost (TPC)	$611,116

Plant Construction Period,	4.0 Years (1 or more)	
Construction Interest Rate,	11.2 %	
Adjustment for Interest and Inflation		$76,713

Total Plant Investment (TPI)	$687,829

Prepaid Royalties		$2,384
Initial Catalyst and Chemical Inventory		$132
Startup Costs		$16,375
Spare Parts		$3,056
Working Capital		$6,922
Land,	200 Acres @ $1500/acre	$300

Total Capital Requirement (TCR)	$716,998
$/kW	1909

ANNUAL OPERATING COSTS

Capacity Factor =   85 %

COST ITEM	QUANTITY	UNIT $ PRICE	ANNUAL COST, K$
Coal (Illinois #6)	3,570 T/D	$24.36 /T	$26,981
Consumable Materials			
Water	3,388 T/D	$0.19 /T	$200
HGCU Sorbent	0.04 T/D	$6,000 /T	$71
Nahcolite	2.3 T/D	$275 /T	$196
Ash/Sorbent Disposal Costs	436 T/D	$8.00 /T	$1,082
Plant Labor			
Oper Labor (incl benef)	15 Men/shift	$34.00 /Hr.	$4,455
Supervision & Clerical			$2,950
Maintenance Costs 2.2%			$13,445
Royalties			$270
Other Operating Costs			$983
Total Operating Costs			**$50,633**
By-Product Credits			
Sulfuric Acid	275.9 T/D	$68.00 /T	$5,820
_____	0.0 T/D	$0.00 /T	$0
_____	0.0 T/D	$0.00 /T	$0
_____	0.0 T/D	$0.00 /T	$0
Total By-Product Credits			**$5,820**
Net Operating Costs			**$44,813**

BASES AND ASSUMPTIONS

A. CAPITAL BASES AND DETAILS

	QUANTITY	UNIT $ PRICE	COST, K$
Initial Cat./Chem. Inventory			
Water	86387 T	$0.19 /T	$16
HGCU Sorbent	17 T	$6,000 /T	$100
Nahcolite	59 T	$275 /T	$16
Total Catalyst and Chemical Inventory			$132
Startup costs			
Plant modifications,	2 % TPI		$13,757
Operating costs			$1,966
Fuel			$652
Total Startup Costs			$16,375
Working capital			
Fuel & Consumables inv	60 days supply		$5,308
By-Product inventory	30 days supply		$563
Direct expenses	30 days		$1,051
Total Working Capital			$6,922

B. ECONOMIC ASSUMPTIONS

Project life	20 Years
Book life	20 Years
Tax life	20 Years
Federal and state income tax rate	38.0 %
Tax depreciation method	MACRS
Investment Tax Credit	0.0 %

Financial structure

Type of Security	% of Total	Current Dollar Cost, %	Ret, %	Constant Dollar Cost, %	Ret, %
Debt	80	9.0	7.2	5.8	4.6
Preferred Stock	0	3.0	0.0	0.0	0.0
Common Stock	20	20.0	4.0	16.5	3.3
Discount rate (cost of capital)			11.2		7.9

Inflation rate, % per year	3.0
Real Escalation rates (over inflation)	
Fuel, % per year	-1.1
Operating & Maintenance, % per year	0.0

C. COST OF ELECTRICITY

The approach to determining the cost of electricity is based upon the
methodology described in the Technical Assessment Guide, published by
the Electric Power Research Institute. The cost of electricity is
stated in terms of 10th year levelized dollars.

	Current $	Constant $
Levelizing Factors		
Capital Carrying Charge, 10th yr	0.179	0.148
Fuel, 10th year	1.091	0.948
Operating & Maintenance, 10th yr	1.151	1.000

Cost of Electricity - Levelized	mills/kWh	mills/kWh
Capital Charges	45.9	38.1
Fuel Costs	10.5	9.1
Consumables	0.6	0.6
Fixed Operating & Maintenance	7.7	6.7
Variable Operating & Maintenance	1.4	1.2
By-product	-2.4	-2.1
Total Cost of Electricity	63.8	53.6

Hybrid Cycles (Turbine / SOFC)

Natural Gas Hybrid Turbine / SOFC Cycle

Natural Gas HAT   **19 MW POWER PLANT**

Total Plant Investment		PROCESS	PROCESS	1st Q 2002 Dollar COST, K$
AREA NO	PLANT SECTION DESCRIPTION	CONT, %	CONT, K$	W/O CONT
15	SOFC Generator Equipment	0	$0	$9,238
15	SOFC Power Conditioning Equipment	0	$0	$2,096
15	Gas Turbine Equipment	0	$0	$4,134
18	Balance of Plant Equipment	0	$0	$5,074

Subtotal, Process Plant Cost	$20,543

Project Management and Engineering Fees	$940
Site Preparation	$431
Overhead and Profit	$5,701

Total Plant Cost (TPC)	$27,615

Spare Parts, Startup, and Land Allowance	$431

Total Capital Requirement (TCR)	$28,046
$/kW	1476

ANNUAL OPERATING COSTS

Capacity Factor =   85 %

COST ITEM	QUANTITY	UNIT $ PRICE	ANNUAL COST, K$
Natural Gas	2,536 1000 SCF/day	$3.20 /1000 SCF	$2,518

Plant Labor

Oper Labor (incl benef)	1 Men/shift	$34.00 /Hr.	$297
Supervision & Clerical			$94

Maintenance Costs $ 0.01 per GT kWe $40

Royalties $0

Other Operating Costs $31

Total Operating Costs			$2,980

By-Product Credits

	0.0 T/D	$0.00 /T	$0
_____	0.0 T/D	$0.00 /T	$0
_____	0.0 T/D	$0.00 /T	$0
_____	0.0 T/D	$0.00 /T	$0

Total By-Product Credits			$0

Net Operating Costs			$2,980

B. ECONOMIC ASSUMPTIONS

Project life	20	Years
Book life	20	Years
Tax life	20	Years
Federal and state income tax rate	38.0 %	
Tax depreciation method	MACRS	
Investment Tax Credit	0.0 %	
Financial structure		

Type of Security	% of Total	Current Dollar Cost, %	Current Dollar Ret, %	Constant Dollar Cost, %	Constant Dollar Ret, %
Debt	80	9.0	7.2	5.8	4.6
Preferred Stock	0	3.0	0.0	0.0	0.0
Common Stock	20	20.0	4.0	16.5	3.3
Discount rate (cost of capital)			11.2		7.9

Inflation rate, % per year	3.0
Real Escalation rates (over inflation)	
Fuel, % per year	0.2
Operating & Maintenance, % per year	0.0

C. COST OF ELECTRICITY

The approach to determining the cost of electricity is based upon the
methodology described in the Technical Assessment Guide, published by
the Electric Power Research Institute. The cost of electricity is
stated in terms of 10th year levelized dollars.

	Current $	Constant $
Levelizing Factors		
Capital Carrying Charge, 10th yr	0.179	0.148
Fuel, 10th year	1.162	1.010
Operating & Maintenance, 10th yr	1.151	1.000

Cost of Electricity - Levelized	mills/kWh	mills/kWh
Capital Charges	35.5	29.4
Fuel Costs	20.7	18.0
Consumables	0.0	0.0
Fixed Operating & Maintenance	3.2	2.8
Variable Operating & Maintenance	3.8	3.3
By-product	0.0	0.0
Total Cost of Electricity	63.1	53.4

Hybrid Cycles (Turbine / SOFC)

Destec (E-Gas^{TM}) / HGCU / "G" GT / No CO_2 Capture

Destec Hybrid HGCU/ SOFC IGCC   **644 MW POWER PLANT**

(no CO2 Capture)   1st Q 2002 Dollar

Total Plant Investment

AREA NO	PLANT SECTION DESCRIPTION	PROCESS CONT, %	PROCESS CONT, K$	COST, K$ W/O CONT
11	Coal Slurry Preparation	0	$0	$32,927
12	Oxygen Plant	0	$0	$60,463
12	Destec Gasifier	5	$3,659	$73,186
12	Misc. Compressors (Recycle, Quench, Air Boost	5	$422	$8,445
14	Gas Conditioning	10	$1,906	$19,061
14	Transport Desulfurizer	15	$1,533	$10,221
14	Sulfuric Acid Plant	0	$0	$23,331
15	Solid Oxide Fuel Cell	0	$0	$177,120
15	Gas Turbine System	5	$2,905	$58,105
15	HRSG/Steam Turbine	5	$2,731	$54,621
18	Water Systems	0	$0	$18,112
30	Civil/Structural/Architectural	0	$0	$32,084
40	Piping	0	$0	$18,112
50	Control/ Instrumentation	0	$0	$20,699
60	Electrical	0	$0	$45,021

Subtotal, Process Plant Cost	$651,509

Engineering Fees	$65,151
Process Contingency (Using cont. listed)	$13,157
Project Contingency, 15 % Proc Plt & Gen Plt Fac	$97,726

Total Plant Cost (TPC)	$827,543

Plant Construction Period,	4.0 Years (1 or more)	
Construction Interest Rate,	11.2 %	
Adjustment for Interest and Inflation		$103,881

Total Plant Investment (TPI)	$931,424

Prepaid Royalties	$3,258
Initial Catalyst and Chemical Inventory	$430
Startup Costs	$21,871
Spare Parts	$4,138
Working Capital	$8,085
Land, 200 Acres	$1,300

Total Capital Requirement (TCR)	$970,505
$/kW	1508

ANNUAL OPERATING COSTS

Capacity Factor = 85 %

COST ITEM	QUANTITY	UNIT $ PRICE	ANNUAL COST, K$
Coal (Illinois #6)	4,145 T/D	$24.36 /T	$31,324
Consumable Materials			
Water	2,931 T/D	$0.19 /T	$173
HGCU Sorbent	0.15 T/D	$6,000 /T	$285
Nahcolite	2.3 T/D	$275 /T	$196
Ash/Sorbent Disposal Costs	617 T/D	$8.00 /T	$1,531
Plant Labor			
Oper Labor (incl benef)	15 Men/shift	$34.00 /Hr.	$4,455
Supervision & Clerical			$3,521
Maintenance Costs 2.2%			$18,206
Royalties			$313
Other Operating Costs			$1,174
Total Operating Costs			**$61,179**

By-Product Credits			
Sulfuric Acid	316.7 T/D	$68.00 /T	$6,681
_____	0.0 T/D	$0.00 /T	$0
_____	0.0 T/D	$0.00 /T	$0
_____	0.0 T/D	$0.00 /T	$0
Total By-Product Credits			$6,681

Net Operating Costs			$54,497

BASES AND ASSUMPTIONS

A. CAPITAL BASES AND DETAILS

	QUANTITY	UNIT $ PRICE	COST, K$
Initial Cat./Chem. Inventory			
Water	74751 T	$0.19 /T	$14
HGCU Sorbent	67 T	$6,000 /T	$399
Nahcolite	59 T	$275 /T	$16
Total Catalyst and Chemical Inventory			$430
Startup costs			
Plant modifications,	2 % TPI		$18,628
Operating costs			$2,486
Fuel			$757
Total Startup Costs			$21,871
Working capital			
Fuel & Consumables inv	60 days supply		$6,184
By-Product inventory	30 days supply		$646
Direct expenses	30 days		$1,254
Total Working Capital			$8,085

B. ECONOMIC ASSUMPTIONS

Project life	20 Years
Book life	20 Years
Tax life	20 Years
Federal and state income tax rate	38.0 %
Tax depreciation method	MACRS
Investment Tax Credit	0.0 %
Financial structure	

Type of Security	% of Total	Current Dollar Cost, %	Ret, %	Constant Dollar Cost, %	Ret, %
Debt	80	9.0	7.2	5.8	4.6
Preferred Stock	0	3.0	0.0	0.0	0.0
Common Stock	20	20.0	4.0	16.5	3.3
Discount rate (cost of capital)			11.2		7.9

Inflation rate, % per year	3.0
Real Escalation rates (over inflation)	
Fuel, % per year	-1.1
Operating & Maintenance, % per year	0.0

C. COST OF ELECTRICITY

The approach to determining the cost of electricity is based upon the methodology described in the Technical Assessment Guide, published by the Electric Power Research Institute. The cost of electricity is stated in terms of 10th year levelized dollars.

	Current $	Constant $
Levelizing Factors		
Capital Carrying Charge, 10th yr	0.179	0.148
Fuel, 10th year	1.091	0.948
Operating & Maintenance, 10th yr	1.151	1.000

Cost of Electricity - Levelized	mills/kWh	mills/kWh
Capital Charges	36.3	30.1
Fuel Costs	7.1	6.2
Consumables	0.5	0.5
Fixed Operating & Maintenance	5.6	4.9
Variable Operating & Maintenance	1.0	0.9
By-product	-1.6	-1.4
Total Cost of Electricity	48.9	41.1

Hybrid Cycles (Turbine / SOFC)

Destec High Pressure (E-Gas^{TM}) / HGCU /
"G" GT / CO$_2$ Capture

Hybrid DESTEC HGCU/ SOFC   **755 MW POWER PLANT**

(Sequesters CO2)   1st Q 2002 Dollar

Total Plant Investment

AREA NO	PLANT SECTION DESCRIPTION	PROCESS CONT, %	PROCESS CONT, K$	COST, K$ W/O CONT
11	Coal Slurry Preparation	0	$0	$40,290
12	Oxygen Plant	0	$0	$109,383
12	Destec Gasifier	5	$4,347	$86,934
12	Misc. Compressors (Recycle, Quench, Air Boost	5	$50	$1,000
14	Gas Conditioning	10	$1,040	$10,400
14	Transport Desulfurizer	15	$1,049	$6,996
14	Sulfuric Acid Plant	0	$0	$28,431
14	Hydrogen Separation Device	50	$4,021	$8,041
15	H2/ CO2 Compressors	0	$0	$52,131
15	Gas Expanders	0	$0	$14,165
15	Solid Oxide Fuel Cell	0	$0	$259,280
15	Gas Turbine System	5	$2,680	$53,595
15	HRSG/Steam Turbine	5	$3,240	$64,798
18	Water Systems	0	$0	$25,741
30	Civil/Structural/Architectural	0	$0	$45,598
40	Piping	0	$0	$25,741
50	Control/ Instrumentation	0	$0	$29,418
60	Electrical	0	$0	$63,984

Subtotal, Process Plant Cost	$925,925

Engineering Fees	$92,592
Process Contingency (Using cont. listed)	$16,426
Project Contingency, 15 % Proc Plt & Gen Plt Fac	$138,889

Total Plant Cost (TPC)	$1,173,833

Plant Construction Period,   4.0 Years (1 or more)

Construction Interest Rate,   11.2 %

Adjustment for Interest and Inflation	$147,351

Total Plant Investment (TPI)	$1,321,183

Prepaid Royalties	$4,630
Initial Catalyst and Chemical Inventory	$195
Startup Costs	$30,701
Spare Parts	$5,869
Working Capital	$10,651
Land, 200 Acres	$1,300

Total Capital Requirement (TCR)	$1,374,529
$/kW	1822

ANNUAL OPERATING COSTS

Capacity Factor =   85 %

COST ITEM	QUANTITY	UNIT $ PRICE	ANNUAL COST, K$
Coal (Illinois #6)	5,530 T/D	$24.36 /T	$41,792
Consumable Materials			
Water	5,059 T/D	$0.19 /T	$298
HGCU Sorbent	0.06 T/D	$6,000 /T	$110
Nahcolite	2.3 T/D	$275 /T	$196
Ash/Sorbent Disposal Costs	832 T/D	$8.00 /T	$2,066
Plant Labor			
Oper Labor (incl benef)	15 Men/shift	$34.00 /Hr.	$4,455
Supervision & Clerical			$4,436
Maintenance Costs 2.2%			$25,824
Royalties			$418
Other Operating Costs			$1,479
Total Operating Costs			$81,074
By-Product Credits			
Sulfuric Acid	427.3 T/D	$68.00 /T	$9,014
_____	0.0 T/D	$0.00 /T	$0
_____	0.0 T/D	$0.00 /T	$0
_____	0.0 T/D	$0.00 /T	$0
Total By-Product Credits			$9,014
Net Operating Costs			$72,060

BASES AND ASSUMPTIONS

A. CAPITAL BASES AND DETAILS

	QUANTITY	UNIT $ PRICE	COST, K$
Initial Cat./Chem. Inventory			
Water	129006 T	$0.19 /T	$25
HGCU Sorbent	26 T	$6,000 /T	$155
Nahcolite	59 T	$275 /T	$16
Total Catalyst and Chemical Inventory			$195
Startup costs			
Plant modifications,	2 % TPI		$26,424
Operating costs			$3,267
Fuel			$1,010
Total Startup Costs			$30,701
Working capital			
Fuel & Consumables inv	60 days supply		$8,199
By-Product inventory	30 days supply		$872
Direct expenses	30 days		$1,580
Total Working Capital			$10,651

B. ECONOMIC ASSUMPTIONS

Project life	20 Years
Book life	20 Years
Tax life	20 Years
Federal and state income tax rate	38.0 %
Tax depreciation method	MACRS
Investment Tax Credit	0.0 %
Financial structure	

Type of Security	% of Total	Current Dollar Cost, %	Ret, %	Constant Dollar Cost, %	Ret, %
Debt	80	9.0	7.2	5.8	4.6
Preferred Stock	0	3.0	0.0	0.0	0.0
Common Stock	20	20.0	4.0	16.5	3.3
Discount rate (cost of capital)			11.2		7.9

Inflation rate, % per year	3.0
Real Escalation rates (over inflation)	
Fuel, % per year	-1.1
Operating & Maintenance, % per year	0.0

C. COST OF ELECTRICITY

The approach to determining the cost of electricity is based upon the methodology described in the Technical Assessment Guide, published by the Electric Power Research Institute. The cost of electricity is stated in terms of 10th year levelized dollars.

	Current $	Constant $
Levelizing Factors		
Capital Carrying Charge, 10th yr	0.179	0.148
Fuel, 10th year	1.091	0.948
Operating & Maintenance, 10th yr	1.151	1.000

Cost of Electricity - Levelized	mills/kWh	mills/kWh
Capital Charges	43.8	36.3
Fuel Costs	8.1	7.1
Consumables	0.5	0.5
Fixed Operating & Maintenance	6.4	5.5
Variable Operating & Maintenance	1.1	1.0
By-product	-1.8	-1.6
Total Cost of Electricity	58.1	48.8

Hybrid Cycles (Turbine / SOFC)

Destec (E-Gas^{TM}) / OTM / CGCU /
"G" GT / No CO_2 Capture

Case: OTM/SOFC Case Destec Cold Gas Cleanup Unit
Plant Size: 675.2 MW
Capacity Factor : 85 %   **1st Quarter 2002 Dollar Base**

Capital Costs		$ x 1000
Installed Equipment Cost		$612,059
Process Contingency		$6,565
Project Contingency		$91,809
Engineering Fees		$61,206
	Subtotal, Process Plant Cost	**$771,639**
AFDC		$96,863
Plant Construction Period	4.0 Years	
Construction Interest Rate	11.2 %	
	Total Plant Investment (TPI)	**$868,502**
Prepaid Royalties		$3,060
Startup Costs		$20,500
Spare Parts		$3,858
Working Capital		$7,836
Land,	200 Acres	$1,300
	Total Capital Requirement (TCR)	**$905,057**
		1340 $/kW

ANNUAL OPERATING COSTS

COST ITEM	Quantity	Unit Price	Annual Cost, K$
Coal (Illinois #6)	4,311 T/D	$24.36 /T	$32,584
Consumable Materials			
Water	5,165 T/D	$0.19 /T	$304
MDEA Solvent	403.2 Lb/D	$1.45 /Lb	$218
Claus Catalyst	0.01 T/D	$470 /T	$2
SCOT Activated Alumina	15.9 Lb/D	$0.67 /Lb	$4
SCOT Cobalt Catalyst			$6
SCOT Chemicals			$19
Ash Disposal Costs	571 T/D	$8.00 /T	$1,417
Plant Labor			
Oper Labor (incl benef)	15 Men/shift	$34.00 /Hr.	$4,455
Supervision & Clerical			$3,374
Maintenance Costs	2.2%		$16,976
Royalties			$326
Other Operating Costs			$1,125
	SubTotal Operating Costs		$60,809
By-Product Credits			
Sulfur	106.0 T/D	$75.00 /T	$2,467
	0.0 T/D	$0.00 /T	$0
	Total By-Product Credits		$2,467
Net Operating Costs			**$58,342**

B-114

CAPITAL BASES AND DETAILS

Startup costs

Plant modifications,	2 % TPI	$17,370
Operating costs		$2,342
Fuel		$788
Total Startup Costs		$20,500

Working capital

Fuel & Consumables inv	60 days supply	$6,396
By-Product inventory	30 days supply	$239
Direct expenses	30 days	$1,202
Total Working Capital		$7,836

ECONOMIC ASSUMPTIONS

Project life	20 Years
Book life	20 Years
Tax life	20 Years
Federal and state income tax rate	38.0 %
Tax depreciation method	ACRS
Investment Tax Credit	0.0 %

Financial structure

Type of Security	% of Total	Current Dollar Cost, %	Ret, %	Constant Dollar Cost, %	Ret, %
Debt	80	9.0	7.2	5.8	4.6
Preferred Stock	0	3.0	0.0	0.0	0.0
Common Stock	20	20.0	4.0	16.5	3.3
Discount rate (cost of capital)			11.2		7.9

Inflation rate, % per year	3.0
Real Escalation rates (over inflation)	
Fuel, % per year	-1.1
Operating & Maintenance, % per year	0.0

COST OF ELECTRICITY

The approach to determining the cost of electricity is based upon the methodology described in the Technical Assessment Guide, published by the Electric Power Research Institute, The cost of electricity is stated in terms of 10th year levelized dollars.

	Current $	Constant $
Levelizing Factors		
Capital Carrying Charge, 10th yr	0.179	0.148
Fuel, 10th year	1.091	0.948
Operating & Maintenance, 10th yr	1.151	1.000

Cost of Electricity - Levelized	mills/kWh	mills/kWh
Capital Charges	32.2	26.7
Fuel Costs	7.1	6.1
Consumables	0.5	0.4
Fixed Operating & Maintenance	5.1	4.4
Variable Operating & Maintenance	0.9	0.8
By-product	-0.6	-0.5
Total Cost of Electricity	45.2	38.0

Humid Air Turbine (HAT)

Natural Gas / Pratt Whitney GT

Natural Gas HAT   **319 MW POWER PLANT**

| Total Plant Investment | | PROCESS | PROCESS | 1st Q 2002 Dollar COST, K$ |
AREA NO	PLANT SECTION DESCRIPTION	CONT, %	CONT, K$	W/O CONT
15	HAT Gas Turbine	10	$8,822	$88,224
15	HAT Heat Recovery	10	$2,399	$23,993
15	HAT Air Saturator	10	$740	$7,402
18	Water Systems	0	$0	$8,493
30	Civil/Structural/Architectural	0	$0	$16,627
40	Piping	0	$0	$8,493
50	Control/ Instrumentation	0	$0	$9,569
60	Electrical	0	$0	$18,900

Subtotal, Process Plant Cost	$181,701

Engineering Fees	$18,170
Process Contingency (Using cont. listed)	$11,962
Project Contingency, 15 % Proc Plt & Gen Plt Fac	$27,255

Total Plant Cost (TPC)	$239,088

Plant Construction Period, 3.0 Years (1 or more)	
Construction Interest Rate, 11.2 %	
Adjustment for Interest and Inflation	$19,486

Total Plant Investment (TPI)	$258,574

Prepaid Royalties	$909
Initial Catalyst and Chemical Inventory	$0
Startup Costs	$7,288
Spare Parts	$1,195
Working Capital	$10,178
Land, 100 Acres @ $1500/acre	$150

Total Capital Requirement (TCR)	$278,293
$/kW	873

ANNUAL OPERATING COSTS

Capacity Factor = 85 %

COST ITEM	QUANTITY	UNIT $ PRICE	ANNUAL COST, K$
Natural Gas	49,802 1000 SCF/day	$3.20 /1000 SCF	$49,443
Consumable Materials			
Water	6,485 T/D	$0.19 /T	$382
Ash/Sorbent Disposal Costs	0 T/D	$8.00 /T	$0
Plant Labor			
Oper Labor (incl benef)	10 Men/shift	$34.00 /Hr.	$2,970
Supervision & Clerical			$1,522
Maintenance Costs	2.2%		$5,260
Royalties			$494
Other Operating Costs			$507
Total Operating Costs			**$60,580**
By-Product Credits			
	0.0 T/D	$0.00 /T	$0
_____	0.0 T/D	$0.00 /T	$0
Total By-Product Credits			$0
Net Operating Costs			**$60,580**

B-119

BASES AND ASSUMPTIONS

A. CAPITAL BASES AND DETAILS

	QUANTITY	UNIT $ PRICE	COST, K$
Initial Cat./Chem. Inventory			
Water	165378 T	$0.19 /T	$31
Total Catalyst and Chemical Inventory			$31
Startup costs			
Plant modifications,	2 % TPI		$5,171
Operating costs			$921
Fuel			$1,195
Total Startup Costs			$7,288
Working capital			
Fuel & Consumables inv	60 days supply		$9,636
By-Product inventory	30 days supply		$0
Direct expenses	30 days		$542
Total Working Capital			$10,178

B. ECONOMIC ASSUMPTIONS

Project life	20 Years
Book life	20 Years
Tax life	20 Years
Federal and state income tax rate	38.0 %
Tax depreciation method	MACRS
Investment Tax Credit	0.0 %
Financial structure	

Type of Security	% of Total	Current Dollar Cost, %	Ret, %	Constant Dollar Cost, %	Ret, %
Debt	80	9.0	7.2	5.8	4.6
Preferred Stock	0	3.0	0.0	0.0	0.0
Common Stock	20	20.0	4.0	16.5	3.3
Discount rate (cost of capital)			11.2		7.9

Inflation rate, % per year	3.0
Real Escalation rates (over inflation)	
Fuel, % per year	0.2
Operating & Maintenance, % per year	0.0

C. COST OF ELECTRICITY

The approach to determining the cost of electricity is based upon the
methodology described in the Technical Assessment Guide, published by
the Electric Power Research Institute. The cost of electricity is
stated in terms of 10th year levelized dollars.

	Current $	Constant $
Levelizing Factors		
Capital Carrying Charge, 10th yr	0.179	0.148
Fuel, 10th year	1.162	1.010
Operating & Maintenance, 10th yr	1.151	1.000

Cost of Electricity - Levelized	mills/kWh	mills/kWh
Capital Charges	21.0	17.4
Fuel Costs	24.2	21.0
Consumables	0.2	0.2
Fixed Operating & Maintenance	4.4	3.9
Variable Operating & Maintenance	5.2	4.5
By-product	0.0	0.0
Total Cost of Electricity	**55.0**	**47.0**

Humid Air Turbine (HAT)

Coal Syngas / Destec (E-Gas^{TM}) / CGCU / Pratt Whitney GT

Destec Coal IGHAT   **407 MW POWER PLANT**

Total Plant Investment

AREA NO	PLANT SECTION DESCRIPTION	PROCESS CONT, %	PROCESS CONT, K$	1st Q 2002 Dollar COST, K$ W/O CONT
11	Coal Slurry Preparation	0	$0	$28,073
12	Oxygen Plant	0	$0	$46,460
12	Destec Gasifier	5	$1,378	$27,555
12	Recycle Gas Compressor	0	$0	$1,914
12	Syngas Cooler/ Fuel Reheater/ Cyclone	0	$0	$3,881
14	Low Temperature Gas Treatment	0	$0	$9,911
14	MDEA/Claus/SCOT	0	$0	$19,785
14	Clean Fuel Compressor	0	$0	$10,936
15	HAT Gas Turbine	10	$10,803	$108,031
15	HAT Heat Recovery	10	$2,770	$27,701
15	HAT Air Saturator	10	$740	$7,405
18	Water Systems	0	$0	$16,041
30	Civil/Structural/Architectural	0	$0	$18,082
40	Piping	0	$0	$16,041
50	Control/ Instrumentation	0	$0	$11,666
60	Electrical	0	$0	$25,374

Subtotal, Process Plant Cost				$378,855

Engineering Fees			$37,886
Process Contingency (Using cont. listed)			$15,691
Project Contingency,	15 % Proc Plt & Gen Plt Fac		$56,828

Total Plant Cost (TPC)	$489,261

Plant Construction Period,	4.0 Years (1 or more)	
Construction Interest Rate,	11.2 %	
Adjustment for Interest and Inflation		$61,417

Total Plant Investment (TPI)	$550,677

Prepaid Royalties			$1,894
Initial Catalyst and Chemical Inventory			$120
Startup Costs			$13,347
Spare Parts			$2,446
Working Capital			$6,131
Land,	200 Acres	@ $1500/acre	$300

Total Capital Requirement (TCR)		$574,915
	$/kW	1411

ANNUAL OPERATING COSTS

Capacity Factor =   85 %

COST ITEM	QUANTITY	UNIT $ PRICE	ANNUAL COST, K$
Coal (Illinois #6)	3,300 T/D	$24.36 /T	$24,942
Consumable Materials			
Water	13,274 T/D	$0.19 /T	$782
MDEA Solvent	403.2 Lb/D	$1.45 /Lb	$181
Claus Catalyst	0.01 T/D	$470 /T	$1
SCOT Activated Alum	15.9 Lb/D	$0.67 /Lb	$3
SCOT Cobalt Catalyst			$5
SCOT Chemicals			$16
Ash/Sorbent Disposal Costs	322 T/D	$8.00 /T	$799
Plant Labor			
Oper Labor (incl benef)	15 Men/shift	$34.00 /Hr.	$4,455
Supervision & Clerical			$2,628
Maintenance Costs	2.2%		$10,764
Royalties			$249
Other Operating Costs			$876
Total Operating Costs			**$45,704**
By-Product Credits			
Sulfur	81.6 T/D	$75.00 /T	$1,899
_____	0.0 T/D	$0.00 /T	$0
_____	0.0 T/D	$0.00 /T	$0
_____	0.0 T/D	$0.00 /T	$0
Total By-Product Credits			$1,899
Net Operating Costs			**$43,804**

BASES AND ASSUMPTIONS

A. CAPITAL BASES AND DETAILS

	QUANTITY	UNIT $ PRICE	COST, K$
Initial Cat./Chem. Inventory			
Water	338490 T	$0.19 /T	$64
MDEA Solvent	10282 Lb	$1.45 /Lb	$15
Claus Catalyst	0.3 T	$470 /T	$0
SCOT Activated Alum	405 Lb	$0.67 /Lb	$0
SCOT Cobalt Catalyst			$16
SCOT Chemicals			$24
Total Catalyst and Chemical Inventory			$120
Startup costs			
Plant modifications,	2 % TPI		$11,014
Operating costs			$1,730
Fuel			$603
Total Startup Costs			$13,347
Working capital			
Fuel & Consumables inv	60 days supply		$5,011
By-Product inventory	30 days supply		$184
Direct expenses	30 days		$936
Total Working Capital			$6,131

B. ECONOMIC ASSUMPTIONS

Project life	20 Years
Book life	20 Years
Tax life	20 Years
Federal and state income tax rate	38.0 %
Tax depreciation method	MACRS
Investment Tax Credit	0.0 %

Financial structure

Type of Security	% of Total	Current Dollar Cost, %	Current Dollar Ret, %	Constant Dollar Cost, %	Constant Dollar Ret, %
Debt	80	9.0	7.2	5.8	4.6
Preferred Stock	0	3.0	0.0	0.0	0.0
Common Stock	20	20.0	4.0	16.5	3.3
Discount rate (cost of capital)			11.2		7.9

Inflation rate, % per year	3.0
Real Escalation rates (over inflation)	
Fuel, % per year	-1.1
Operating & Maintenance, % per year	0.0

C. COST OF ELECTRICITY

The approach to determining the cost of electricity is based upon the
methodology described in the Technical Assessment Guide, published by
the Electric Power Research Institute. The cost of electricity is
stated in terms of 10th year levelized dollars.

	Current $	Constant $
Levelizing Factors		
Capital Carrying Charge, 10th yr	0.179	0.148
Fuel, 10th year	1.091	0.948
Operating & Maintenance, 10th yr	1.151	1.000

Cost of Electricity - Levelized	mills/kWh	mills/kWh
Capital Charges	33.9	28.1
Fuel Costs	9.0	7.8
Consumables	0.7	0.6
Fixed Operating & Maintenance	6.1	5.3
Variable Operating & Maintenance	1.1	0.9
By-product	-0.7	-0.6
Total Cost of Electricity	**50.1**	**42.1**

Appendix C - FUEL COMPOSITION

Ambient conditions:

Temperature	59 F
Pressure	14.7 psia
Relative Humidity	60%

Coal Analysis

Proximate Analysis	(Wt. %)	(Wt. % dry)	Ultimate Analysis	(Wt. %)	(Wt. % dry)
Moisture	11.12		Moisture	11.12	
Ash	9.70	10.91	Carbon	63.75	71.72
Volatiles	34.99	39.37	Hydrogen	4.50	5.06
Fixed carbon	44.19	49.72	Nitrogen	1.25	1.41
Total	100.00	100.00	Chlorine	0.29	0.33
			Sulfur	2.51	2.82
HHV (Btu/lb)	11,666	13,126	Ash	9.70	10.91
			Oxygen	6.88	7.75
			Total	100.00	100.00

NATURAL GAS – assumed 100% Methane for ASPEN simulation.

Goals

The primary goal of the Vision 21 Program is to effectively remove all environmental concerns associated with the use of fossil fuels for producing electricity, transportation fuels, and high-value chemicals. This goal is to be accomplished at competitive costs. The specific performance targets, costs, and timing for Vision 21 plants are shown below.

Vision 21 Energy Plant Performance Targets

Efficiency - Electricity Generation:

- 60% for coal-based systems (HHV)
- 75% for natural gas-based systems (LHV)

Efficiency - Combined Heat & Power:

- Overall thermal efficiency above 85% (HHV); also meets efficiency goals for electricity (based on fuel)

Efficiency - Fuels Only Plant:

- 75% feedstock utilization efficiency (LHV) when producing fuels such as H_2 or liquid transportation fuels alone from coal

Environmental:

- Atmospheric release of near zero emissions of
  - sulfur
  - nitrogen oxides
  - particulate matter
  - trace elements and organic compounds or liquid transportation fuels alone from coal
- 40-50% reduction of CO_2 emissions by efficiency improvement
  - 100% reduction with sequestration

Costs:

- Aggressive targets for capital and operating costs and RAM (reliability, availability, and maintenance). Cost of electricity 10% lower than conventional systems
- Products of Vision 21 plants must be cost-competitive with other energy subsystems with comparable environmental performance, including specific carbon emissions

Timing:

- Major benefits from improved technologies begin by 2005
- Designs for most Vision 21 subsystems and modules available by 2012
- Vision 21 commercial plant designs available by 2015